▶ **DiY WiFi**

DOI: 10.1057/9781137312532.0001

Other Palgrave Pivot titles

Majid Yar: **Crime, Deviance and Doping: Fallen Sports Stars, Autobiography and the Management of Stigma**

Grace Ji-Sun Kim and Jenny Daggers: **Reimagining with Christian Doctrines: Responding to Global Gender Injustices**

Love Henry Whelchel: **Sherman's March and the Emergence of the Independent Black Church Movement: From Atlanta to the Sea to Emancipation**

G. Douglas Atkins: **Swift, Joyce, and the Flight from Home: Quests of Transcendence and the Sin of Separation**

David Beer: **Punk Sociology**

Owen Anderson: **Reason and Faith in the Theology of Charles Hodge: American Common Sense Realism**

Jenny Ruth Ritchie and Mere Skerrett: **Early Childhood Education in Aotearoa New Zealand: History, Pedagogy, and Liberation**

Pasquale Ferrara: **Global Religions and International Relations: A Diplomatic Perspective**

François Bouchetoux: **Writing Anthropology: A Call for Uninhibited Methods**

Robin M. Lauermann: **Constituent Perceptions of Political Representation: How Citizens Evaluate their Representatives**

Erik Eriksen: **The Normativity of the European Union**

Jeffery Burds: **Holocaust in Rovno: A Massacre in Ukraine, November 1941**

Timothy Messer-Kruse: **Tycoons, Scorchers, and Outlaws: The Class War That Shaped American Auto Racing**

Ofelia García and Li Wei: **Translanguaging: Language, Bilingualism and Education**

Øyvind Eggen and Kjell Roland: **Western Aid at a Crossroads: The End of Paternalism**

Roberto Roccu: **The Political Economy of the Egyptian Revolution: Mubarak, Economic Reforms and Failed Hegemony**

Stephanie Stone Horton: **Affective Disorder and the Writing Life: The Melancholic Muse**

Barry Stocker: **Kierkegaard on Politics**

Michael J. Osborne: **Multiple Interest Rate Analysis: Theory and Applications**

Lauri Rapeli: **The Conception of Citizen Knowledge in Democratic Theory**

Michele Acuto and Simon Curtis: **Reassembling International Theory: Assemblage Thinking and International Relations**

Stephan Klingebiel: **Development Cooperation: Challenges of the New Aid Architecture**

Mia Moody-Ramirez and Jannette Dates: **The Obamas and Mass Media: Race, Gender, Religion, and Politics**

Kenneth Weisbrode: **Old Diplomacy Revisited**

Christopher Mitchell: **Decentralization and Party Politics in the Dominican Republic**

palgrave▸pivot

DiY WiFi: Re-imagining Connectivity

Katrina Jungnickel
Goldsmith, University of London, UK

DOI: 10.1057/9781137312532.0001

First published 2014 by
PALGRAVE MACMILLAN

Palgrave Macmillan in the UK is an imprint of Macmillan Publishers Limited, registered in England, company number 785998, of Houndmills, Basingstoke, Hampshire RG21 6XS.

Palgrave Macmillan in the US is a division of St Martin's Press LLC, 175 Fifth Avenue, New York, NY 10010.

Palgrave Macmillan is the global academic imprint of the above companies and has companies and representatives throughout the world.

Palgrave® and Macmillan® are registered trademarks in the United States, the United Kingdom, Europe and other countries

ISBN: 978–1–137–31254–9 EPUB
ISBN: 978–1–137–31253–2 PDF
ISBN: 978–1–137–31252–5 Hardback

This book is printed on paper suitable for recycling and made from fully managed and sustained forest sources. Logging, pulping and manufacturing processes are expected to conform to the environmental regulations of the country of origin.

A catalogue record for this book is available from the British Library.

A catalog record for this book is available from the Library of Congress.

www.palgrave.com/pivot

DOI: 10.1057/9781137312532

I like making things. I'm interested in what makes people cooperate. I'm passionate about people trying things, making mistakes, learning, trying again and doing it. Not saying: 'You can't do that'. Not saying: 'Scientists do that' or 'Companies do that' or 'Somebody does that'. Don't think that way. Give it a go. Try it out. Build a car. Build a rocket. Build whatever. Give it a go. Try and learn. And the bonds that develop ... I've seen so many friendships made and people go on and pursue careers from the experience and get into different industries that they probably never thought about ... And as you learn and you go: 'Oh, look that wasn't so hard'. 'Oh that wasn't so scary'. I bet you never thought you'd be writing a book. Bet you thought, 'Oh no way could I do that'. I think it's one of the problems with society, as we become consumers we say, 'No, I couldn't do that'. Why can't you do it?

Ron
Adelaide, South Australia

DOI: 10.1057/9781137312532.0001

Contents

DOI: 10.1057/9781137312532.0002

List of Figures

Acknowledgements

This book began as a doctoral thesis at Studio INCITE (Incubator for Critical Inquiry into Technology and Ethnography) in the Sociology Department at Goldsmiths College, University of London. While writing can at times feel like a lonely pursuit, I was never alone. Now, in its final form, this book veritably hums with the collaborative energy of a community of people who have contributed, participated and keenly shaped my work. My thanks go to Nina Wakeford and Genevieve Bell who have been instrumental in helping to press my ideas into robust sociological arguments and me into a sociologist, encouraging and inspiring me in my scholarly endeavours for many years. I owe much to the WiFi community – Robert, Hat, Lucychili, Adhoc, Leafkicker, Seabird, Karl, Cougar, Shadey, LoC, Duncanj, DJ, Didz, DrGeforce3, Rids, Paul and everyone else at Air-Stream – for showing interest in my research, spending time with me, teaching me many things and taking me places I would never have gone on my own. Your continued support and friendship are highly valued. I also extend my heartfelt thanks to James Fraser, Ben Bowering, Mel Gregg, Lara Houston, Julien McHardy, Amanda Windle, Britt Hatzius, Amber Templemore-Finlayson, Tillie Harris, Simon Cohn and my parents, Raylee and Gunther, for the many random acts of thoughtfulness and indefatigable encouragement that cumulatively made the research for this book possible.

I also wish to express my appreciation to Gerard Goggin in the Department of Media and Communications, University of Sydney, for supporting my application for a Visiting Research Fellowship that greatly assisted in finalising the manuscript.

DOI: 10.1057/9781137312532.0004

List of Abbreviations

ACMA   Australian Communications and Media Authority
ADSL   Asymmetric Digital Subscriber Line
ANT   Actor Network Theory
AP   Access Point, also known as node or antenna
DiT   Do-it-Together
DiY   Do-it-Yourself
DNS   Domain Name System
FTP   File Transfer Protocol
ICT   Information and Communication Technology
IM   Instant Messenger
IP   Internet Protocol
IRC   Internet Relay Chat
ISP   Internet Service Provider
Kbps   Kilobits per second
LAN   Local Area Network
LoS   Line of Sight
Mbps   Megabits per second
OSS   Open Source Software
SMS   Short Message Service
VoIP   Voice over Internet Protocol
WiFi   Wireless Fidelity
WIKI   What I Know Is, collaborative editing OSS
WLAN   Wireless Local Area Network

DOI: 10.1057/9781137312532.0005

palgrave▸**pivot**

www.palgrave.com/pivot

1
Introduction

Abstract: *This chapter opens with an account of the theft of a core antenna from a community wireless network. This sets the scene for a study of WiFi that is socially, materially and culturally embedded in a specific place and made by a group of individuals who collectively re-inscribe broadband technology with new meanings and re-imagined possibilities of use. I outline cores themes emerging throughout the book and briefly introduce the chapters that follow.*

Jungnickel, Katrina. *DiY WiFi: Re-imagining Connectivity*. Basingstoke: Palgrave Macmillan, 2014. DOI: 10.1057/9781137312532.0006.

It is a hot Saturday afternoon in March. At 3pm, the heat rises thickly from the tarmac and the eucalyptus trees shimmer in the sun. I am seated on a short wooden bench near a stone building in the playground of a Central Adelaide public school, the site of weekly meetings for the largest community Wireless Fidelity (WiFi) group in Australia, Air-Stream. Members of the group use WiFi to make their own wireless broadband network that spans the largely suburban city by connecting homemade antennas, many of which are located in their backyards. They use the network to chat, send messages, share files, play games and host blogs and websites. After corresponding for two months I had arranged to meet two members. Tim founded the group with 'a few mates' five years ago in 2001 and has continued to lead it around the demands of his university Information and Communication Technology (ICT) studies. His inquisitive manner and easy confidence belie his 26 years. Ron, 45, is warm and generous in his welcome with a similar easy, open manner. A self-confessed 'PR person' and official secretary, Ron was often the first to respond to media enquiries and questions from new people, his social skills finely honed from his day job as director of a company that provides employment for people with disabilities.

A car pulls into the dusty car park and Tim and Ron quickly cross the playground laden with toolboxes, an antenna and coils of wire. After quick greetings they launch into a dramatic story of theft. The night before, a major antenna in the community network was stolen from the top of a factory in the southeast of the city. At a cost of $5000 it is a significant loss for the group and it is not the first time it has happened. Equipment has been stolen from four different sites with three thefts this year. The first, in June the year before, was on the roof of a large supermarket. Someone cut through the base of the four-metre steel mast to remove two dishes. Then, in February, a wireless box went missing from the roof of a house. The owners were away when it was taken and it was only when Dan, another member of the group, failed to connect to the network and went to check the site that it was reported stolen. Last night's theft is the second this month. A week prior, someone cut the cables and removed the mast from the backyard of a member's home. Overall, the group lost 4 of the 23 major antennas making up the core infrastructure that spans the suburban city.

Tim and Ron are clearly concerned about the recent loss and increasingly worried about the emerging pattern. They think it is possible that thieves are using the many maps, photos and diagrams available on the

DOI: 10.1057/9781137312532.0006

group's website to locate sites and plan attacks. The website is open to the public, to recruit new members and share knowledge. Without this information people would not know where the network was located or how it was put together. With it, there is a chance, and increasing reality, that others might be using it to steal equipment.

Tim is tired. He spent the night on the factory roof attempting to install night vision cameras, but it became too dark, windy and dangerous to continue. When this did not work he developed a unique monitoring system, involving his mobile phone, two computers and quick programming, which would send him a mobile text message should the connection break again. Once alerted, he could race to the site. Tim acknowledged it was a temporary solution. Although it worked, there were several issues including the fact he received a text message every time a bird sat on the antenna.

I was struck by the complexity of WiFi. Like many I used WiFi to access the internet at home, at work and in places in between such as cafes and libraries. I was used to thinking of it as a channel through which I could study, socialise and shop online without wires. Prior to learning about community WiFi networks I had not given much thought to WiFi as something that could be made or for that matter, stolen. Nor had I considered the possibility of different versions or the implications this had for understanding the internet, local technological cultures or practices of innovation. I began to ask: How can you make WiFi? Do different makers matter? What would alternate versions of WiFi look like?

WiFi may be an electromagnetic radio signal invisible to the human eye but in this account it is present on rooftops, backyards and school playgrounds, on weekends and evenings, in photographs, maps, websites and dramatic stories. It involves birdlife, thieves, improvised methods and a constellation of bought, found and re-purposed materials such as sticky tape, mobile phones and cameras. At a time when broadband and WiFi have become synonymous with the internet, this kind of wireless work unsettles familiar understandings of the point and purpose of conventional pay-for-service connectivity. From a sociological perspective what is remarkable is the sheer array of stuff, people and places involved in making (and re-making) WiFi. Here, information is socialised, materialised and visualised. WiFi is a thing that can be studied. As one member described it: 'We are building Ournet, not the internet'. Put another way, rather than simply adding content *to the* internet, Air-Stream members

DOI: 10.1057/9781137312532.0006

are making the very architecture *of the* internet their own, and in the process re-imagining ways to connect to one another.

Despite running successfully for more than a decade, surprisingly little is known of how or why these networks are made. Drawing on original research using ethnographic methods this book presents an overdue account of community wireless broadband culture. The chapters that follow describe how individual makers, or what I also term 'backyard technologists', collectively make their own WiFi network using a diverse range of at-hand materials and improvised methods in everyday places. Just as early Science and Technology Studies (STS) researchers brought science 'down to earth' (Law and Mol 2001:2), the book illustrates how a highly sophisticated technology, traditionally shaped and controlled by large scale ICT organisations, is made for not-for-profit purpose from the ground up, or in this case, from the backyard out.

The research builds on the idea established in STS and cultural studies that ICTs are not ubiquitous or universal (Miller and Slater 2000; Goggin 2004, 2007; Ito et al. 2005; Goggin and Gregg 2007; Burrell 2009). They do not all follow the same development or adoption trajectories but are constantly being made (and remade) in relation to their material, social and cultural contexts. WiFi presents an intriguing subject due to the 'always on', 'anywhere' and 'everywhere' rhetoric that has surrounded it for the last decade. It is often overlooked (and under-explored) because it is largely considered an invisible and 'boring' infrastructure that simply provides wireless access to the internet. Yet these makers, in building networks from scratch, challenge conventions of what technological artefacts are meant to do and look like. A central tenet of the book is that different versions of WiFi become visible if attention is paid to distinct technology cultures, and in particular those of backyard technologists. Paying attention to local history and its distinctive relationship to technology use/misuse and understanding provides a rich description of a local version of WiFi – an Australian WiFi – and in the process signals the possibility of comparative studies. Ultimately, in drawing attention to different ways of thinking about WiFi development and use in other places, it contributes to a deeper understanding of global wireless digital cultures.

The relevance of a small vivid example like Air-Stream lies in how it renders visible largely invisible digital technologies, pointing out other shapes and possibilities of use and asking new questions about things we take for granted. They make us wonder how they 'might have been

DOI: 10.1057/9781137312532.0006

otherwise' (Bijker and Law 1992:3). WiFi makers shift the register – not by asking what we can do *on* the internet, but what we can do *with* it. Their practice signals ways of connecting with each other that circumvents familiar telecommunication relationships. Stepping outside conventional dependencies transforms our view of the technological landscape. It changes the questions: What else can WiFi do besides pipe the internet? Why is the internet packaged in the way it is? Why should we be content with fast downloads and slow uploads? What does this inhibit? How might things be different? Critically, the study also addresses how this technology is made. WiFi makers imbue a Do-it-Yourself (DiY) ethic, yet they do not do it alone – they Do-it-Together (DiT). This timely critique of collective DiT innovation in an increasingly networked society will be of interest to STS scholars and practitioners of maker culture.

A short note about what this book is not about. It does not seek to be representative of all community wireless technology networks. In fact it does not even claim to represent Air-Stream now. Rather, the book captures and tells a particular story of a particular point in time: 2006–2009. Further to arguing that WiFi is local and cultural, it is also temporal. The book is, and will always be, set against a backdrop of constant technological change and debate about the role and importance of wireless broadband provision and use around the world. Since my fieldwork, the Labour Government started to roll out the Australian National Broadband Network (NBN) with the aim of connecting every citizen to high speed internet. In 2013, the federal election resulted in a new Liberal Government with its own vision of broadband in this country, which again looks to significantly transform how citizens get connected. Irrespective of the scale and nature of new technological infrastructure, these kinds of advancements are all part of and respond to the same set of challenges and critiques. The significance and innovation of the case study in this book lies in the way it interweaves technological imaginaries with local everyday use in order to generate other kinds of futures. It presents a complementary, yet strikingly different narrative to top down technology innovation processes that produce services for consumers. In this way it is a timely, and timeless, reminder of the importance and value of both stories, especially those often rendered invisible, and the evolving relationships between them.

The next two chapters theoretically and methodologically frame the study. In Chapter 2, I develop a framework for exploring WiFi in relation to social, cultural and material ecologies, to argue that marginal

DOI: 10.1057/9781137312532.0006

and mundane technology stories matter. I outline three main themes: the nature of connectivity, the visual culture of a digital technology and DiY/ DiT practice. Chapter 3 begins by introducing the group, location and key characters. I discuss the research design and highlight some of the epistemological, methodological and practical issues that shaped fieldwork. The chapters that follow draw on ethnographic material to tell stories about community WiFi makers. Chapter 4 is located at a monthly WiFi meeting where I examine the multi-dimensional, contrasting and sometimes contradictory nature of connectivity and propose that the group's seemingly scattergun visual culture is designed to fit the idiosyncrasies of the network. I also introduce and explain the role of the barbeque or 'barbie' in the making of WiFi, arguing that this too is enrolled by the group as a means of contending with the complexities of the technology. In Chapter 5, I discuss how WiFi makers deal with uncertainty. Because the network is located across the city, rather than in isolated hotspots, makers encounter a vast array of interruptions on a daily basis in the form of trees, birds, bugs, technical complications, materials and the weather. Rather than attempting to tidy up or erase interruptions, I describe how they build them into the network. Chapter 6 is situated in a suburban yard and rooftop. Describing the process of 'stumbling' for wireless noise reveals how makers represent the digital landscape, use materials at hand and weave a technological imaginary into a network that is never fully stabilized or known. Chapter 7 examines the role of 'mods' (modifications) in digital tinkering practices during the raising of an antenna on a backyard shed. Here, makers render visible mistakes and tangents, thereby reworking conventional spectrums of success and failure. In Chapter 8, I draw on encounters between local internet service providers (ISPs) and WiFi makers to discuss 'homebrew high-tech', a distinctive cultural way of imagining and making a version of wireless broadband that marries precision and tinkering with a collaborative social approach and intimate material knowledge in mundane locations. To conclude, Chapter 9 draws these findings together via the three themes to reflect on (and project futures of) collective DiT technology making cultures.

DOI: 10.1057/9781137312532.0006

2
Who Makes WiFi (and Why Other Makers Matter)?

Abstract: *WiFi is often understood (and overlooked) as a one-size-fits-all phenomenon that exists 'everywhere' and 'anytime' and is packaged in a pay-plug-and-play format by large scale telecommunication distributors, mainly as a way to access the internet without wires. Yet, as this chapter illustrates, it is not ubiquitous or universal. In this case it is uniquely customised, culturally shaped, comprised of ordinary stuff in everyday places and made (and remade) by individuals on a daily basis. Drawing on Science and Technology Studies, I argue that the global starts with the local. In other words, other makers matter.*

Jungnickel, Katrina. *DiY WiFi: Re-imagining Connectivity*. Basingstoke: Palgrave Macmillan, 2014. DOI: 10.1057/9781137312532.0007.

Setting the scene: top down technological trajectories

> We don't have an information super highway – we've got an IT goat track
>
> Kim Beazley quoted in Johnson 2008:3

> Rupert Murdoch thought the situation a 'disgrace'. James Packer said it was 'embarrassing'. Fairfax's David Kirk talked about 'fraudband'
>
> ALP, 2007

> Consultant Mark Pesce said Australia was 'basically an internet backwater'... Broadband is merely the latest chapter in a very old story.
>
> Given 2008:6

Since the mid-1990s the internet has become a significant component of a nation's international, economic and cultural standing in the world and a site for increasingly complicated political discourse. Allen and Long explain how in Australia governmental campaigns in 1995 associated the internet with national identity. 'These campaigns took the form of trying to convince Australians to use the internet – that to be 'Australian' meant getting online – and thus linking citizenship to internet use' (2004:232). One of the biggest challenges to this vision however, has been an inability to adequately service rural and remote parts of a large country and address low speeds and restricted download issues in city centres – a dire situation that many argued has been the legacy of a telecommunications monopoly (Meikle 2004; Rennie and Young 2004). Over a decade later, digital connectivity was a pivotal platform in the 2007 national elections where Kevin Rudd, then prime minister, announced plans to 'revolutionise Australia's internet infrastructure', leaving little ambiguity about its role and importance in Labour's broadband policy (ALP 2007).

From 2010, the rollout of the federally funded NBN was designed to continue this trajectory with the aim of delivering all Australians into a new world of communication. Expected to grow the economy, enhance education and business and expand the quality and quantity of jobs, it was, according to the Labour government that conceived it, nothing short of a 'historic act of nation building' (Rudd 2009). Broadband was again a major platform in the 2013 federal election with fierce debates about how *high-speed* was high-speed broadband for Australian citizens

DOI: 10.1057/9781137312532.0007

(Wardell 2013). Whereas once Australia was largely dependent on the wool industry, with an economy that 'rode on the sheeps' back' (White 1981:149), connectivity via the internet is central to its political and economic present and future.

My intention is not to explain how this situation has arisen or predict large-scale broadband internet futures but to highlight what is present and more importantly absent. What is implicit in these accounts is a top down technologically deterministic pressure to connect coupled with a sense of shame and embarrassment of being left behind in a global context. The history of digital technological infrastructure is in many ways driven by an 'imperative to connect' (Green and Harvey 1999; Green 2000; Green et al. 2005) which Green explains is: '[T]he urgent political and commercial insistence that everyone must connect to ICTs, and must do it now, is heavily loaded with this idea of getting somewhere' (2000:1). Although written over a decade ago this research holds relevance in a contemporary technological landscape because it 'focuses on the connection itself, rather than what it is to be connected or why' and 'leaves out the question of what *disconnections* are entailed in connecting' (Green and Harvey 1999:12; emphasis in original). Paired with the intense drive to get constituents online is a narrow idea of who is responsible and what is considered to be connectivity. Moreover, in the rush to connect, the infrastructure of pay-plug-and-play internet in many ways mirrors that of other services such as electricity, water, television and so on which *push* goods to terminal connections located in homes and offices. Pipes and wires built into the domestic infrastructure shape practices of use. We know how to purchase and use them. We know whom to call when they break. We also know not to touch them. They reflect and reinforce familiar patterns of consumption. Given that broadband internet is most often talked about in terms of download speeds it appears similarly configured. In 2013, Australian average broadband download speeds were up to six times faster than upload speeds – a discrepancy that reinforces the idea of connectivity as something to consume than to contribute.[1] This bi-directionality is not being addressed with the new NBN[2] and is an imbalance keenly felt by all users and especially businesses as explained by this IT journalist:

> The limited upload speed is of particular concern...the answer, often, was to back up a brand-office server to tape, then physically courier the tape to a capital-city office for archiving. (Braue 2010)

DOI: 10.1057/9781137312532.0007

The internet might be ever increasingly quicker and easier to access, yet models of use in Australia remain restrictive, with unequal up/download speeds and consumers locked into long-term expensive bundled contracts (up to 24 months).[3] Even the moniker 'superhighway' lends itself to a particular form of travel, vigorously bound by rules and regulations which do not accommodate a vision of group-facilitated collaboration or creative production. Less acknowledged is the possibility of other means of getting connected, or awareness of other technology makers and versions of the internet that unsettle this stable top down pay-plug-and-play model. The internet in Australia is dominated and therefore shaped by commercial models of government and large scale technology organisations to the point where alternate technology practices do not 'figure much in the commercial realm or in policy making' (Goggin 2007:122). What a study of a community wireless network offers is a way to consider a more individually owned yet collectively operated means of connecting to services and each other.

Why WiFi?

Since its launch in 1998, WiFi[4] has achieved widespread popularity, primarily as a result of being employed by large-scale telecommunication organisations as a means of wirelessly distributing the internet. Yet, it is more than this. WiFi is an electromagnetic radio signal that enables high-speed wireless exchange of data. It provides a way to connect independent computer devices together to facilitate file sharing, Instant Messaging (IM), email, Voice-over Internet Protocol (VoIP) and multi-player gaming. Dependent on the weather, building materials and strength of the original signal, a modern WiFi device equipped with a directional antenna can transmit up to 100 megabits of data per second (Mbps) over distances up to eight kilometres.

Over the past decade, the rhetoric surrounding WiFi has promised a radical departure from everyday traditional fixed line computing practices, seemingly unbound by normative time and space. Liberal use of the words 'freedom', 'flexibility' and 'convenience' frequented descriptions of WiFi in media and commercial discourse (Libbenga 2003; Koprowski 2004). Advertisements, such as those by *Telstra*, Australia's largest telecommunications organisation, featuring images of an outback roadtrip with the tagline 'Wireless is freedom', is a typical example of the

DOI: 10.1057/9781137312532.0007

'anytime', 'everywhere' and 'always on' pay-for-service consumer model. WiFi in reality is not quite as easy to use or access in many places, yet the gist of these messages captured global attention. WiFi-enabled mobile phones, laptops, gaming consoles, cameras and printers have become the norm. Correspondingly, WiFi 'hotspots' where people connect to the internet via a WiFi network can be found not only in homes and workplaces but increasingly in parks, cafes, churches, pubs, trains, planes, airports, surfboards, remote villages and even large stretches of road and beach.[5] However popular and widespread, these commercial models do not account for all WiFi infrastructures in operation. There are many others who make their own WiFi networks, located on the fringes of established formal institutions.

Volunteer not-for-profit community groups all over the world became early adopters of WiFi because unlike other wireless technologies such as mobile telephony it could be designed and shaped for specific local use. WiFi operates on an unlicensed spectrum,[6] which grants individuals the same rights to broadcast and receive wireless signals as corporate and governmental organisations. In Australia, this is called the 'Public Park Concept' and is relatively unregulated by the Australian Communications and Media Authority (ACMA). While ISPs need a Carrier Licence to operate (and sell the internet), not-for-profit community WiFi groups do not guarantee a continuity of service and therefore a license is not required. While this means there are fewer restrictions in gaining entry to this technology landscape, there is also less known about WiFi makers. WiFi has not achieved the attention of similar community based technical hobbies such as ham radio.[7] There are no firm figures for the total number of community WiFi groups around the world but estimates suggest there are at least 400, with the majority located in Europe (113) and North America (48) (Personal Telco 2012). Australia has six, with the largest being in Adelaide. Along with Open Source Software (OSS), which is free to circulate and use without a license, WiFi makes it possible to build alternate communication infrastructures that circumnavigate traditional power relationships. Individuals are able to create their own customised computer networks and avoid the costly charges imposed by ISPs.

RON: You can't run a [phone] cable down the street. No one is going to let you do that. You can't set up your own satellite. You can't do those kinds of thing. Yet all of a sudden this technology allowed average people to set up a communications network themselves and I think that's what captured my attention ... You

DOI: 10.1057/9781137312532.0007

don't have to wait for someone else to come along and you have to pay for it. You can do it yourself.[8]

The popularity of WiFi in the early 2000s emerged in part in relation to excitement about Web 2.0, which promised a paradigm shift in the nature of interaction on the internet. This entailed a move away from a focus on individual consumption, characterised by the first version of the internet (Web 1.0), to one that privileged user-led content generation and participation. The latter reflected Tim Berners-Lee's original concept of an internet that was not just for readers, but for writers (and makers) as well. It was a time that heralded great productive excitement via websites, blogs, social media, gaming, photo sharing and early video editing software. However, it was not only Do-it-Yourself (DiY) activity *on the internet* that was making an impact and changing the way we were thinking about new technologies. WiFi signalled significant DiY activity surrounding the infrastructure *of the internet*. While only a few people attempt to re-engineer telephony (with exception of ham radio operators) or the television, hundreds of volunteer community groups all over the world became united by an interest in collectively remaking wireless infrastructures (Sandvig 2004; Mackenzie 2005a, 2005b, 2007; Forlano 2008; van Oost et al. 2008; Söderberg 2011; Jungnickel 2013).

Community WiFi makers and other stories

Often called enthusiasts or hobbyists, WiFi makers experiment with, build and maintain their own local wireless networks around the demands of salaried employment, family and social commitments in non-conventional contexts such as backyards and sheds (Figure 2.1). They incorporate improvised methods and an array of found, bought and re-appropriated materials. Although there are many organisations that make WiFi, some gain more attention than others. Very rarely, for instance, do representations of community wireless activities attract national interest. Goggin and Gregg suggest that this is partly due to stubborn 'perceptions of wireless as being the domain of the boardroom, the café or the inner-city minimalist apartment' (2007:45). Much like the history of the internet in Australia, wireless technology is largely understood as something to purchase and consume. It is most often represented in terms of top-down 'competitive discourse' and 'nation building exercises such as railways and roads' and 'embedded in terms of responsibility and efficacy' (41).

DOI: 10.1057/9781137312532.0007

FIGURE 2.1   *Julie's WiFi antenna located in her backyard*

This book presents a different view. It attends to what Bijker has argued about how the 'stories we tell about technology reflect and can also affect our understanding of the place of technology in our lives and our society' (1995:1). Given the loudest stories amplified by westernised media tend to represent the point of view of large ISPs and governmental policy makers, what a study of community WiFi makers offers is a way of seeing other shapes and uses of this technology.

> Most of the time, most of us take our technologies for granted. These work more or less adequately, so we don't inquire about why or how it is they work. We don't inquire about the design decisions that shape our artifacts. We don't think very much about the ways in which professional, political, or economic factors may have given form to those designs – or the way in which they were implemented in practice. And even when our technologies go wrong, typically our first instinct is to call the repairperson. There are

DOI: 10.1057/9781137312532.0007

routine methods for putting them right: it doesn't occur to us to inquire deep into their provenance. (Bijker and Law 1992:1–2)

This book documents a group of makers that do not accept a digital technology as it is packaged and purchased for a specific use. They look beyond its designed capabilities, pushing and extending its potential in new contexts. Their approach presents an opportunity to ask questions about existing WiFi systems and structures, to imagine how WiFi 'might have been otherwise' (3). To give a sense of the stories that follow, this is how one member explains how she imagines the role of Air-Stream in the broader technological landscape:

> JAN: We've reached a point where broadcast media, broadcast product, broadcast lots of things are not a good fit anymore because the people who provide the broadcast have become too interested in generating profit out of this model to the extent that it no longer serves the people who receive that service. So people are interested in distributed ways of participating with each other to generate value. It's like we've got freeways and they work which is great but people all have backyards and driveways and it's like Air-Stream is more like everybody linking up their driveways to make a freeway.

Jan's analogy evokes a powerful image of a collective multi-directional community WiFi network in contrast to conventional telecommunication systems that predicate a largely one-way service. The old broadcast model is the freeway. According to Jan, this traditional monolithic and highly regulated distribution model is not the only way people can connect to one another. She does not want to replace the freeways. They have a purpose. Instead, she suggests harnessing resolutely suburban and easily accessible things – backyards and driveways – to explore different ways of encountering, navigating and most importantly contributing to a collaborative infrastructure. These mundane spaces and practices bring WiFi down to earth, embedding it in a familiar everyday landscape and the hands of makers who materially imagine other technical possibilities. This and stories that follow signal a way of attending to and possibly reversing 'the usual binary between commercial/ public versus dominant/private use of communications technology' (Goggin 2007:121).

The study of mundane and marginal technologies

In the context of large commercial telecommunications providers the accomplishments of a small wireless network might appear trivial and

DOI: 10.1057/9781137312532.0007

inconsequential. Yet, the study of a community WiFi network draws attention to the dynamics and dimensions of a technological infrastructure in the making, affording a way of seeing into the 'black box'. This involves gaining entry into the inner workings of an artefact or system, where we can interrogate seemingly closed systems for their sociocultural, gendered, historical and material composition, and ask why we 'get the technologies we deserve' and how and in what ways they 'mirror our societies' (Bijker and Law 1992:3). STS scholars have argued that once paths of innovation and use become established they are harder to change than when they are fresh and new. It is not long before they appear as if they have always been there, 'characterized by perfect order, completeness, immanence and internal homogeneity rather than leaky, partial and heterogeneous entities' (Graham and Thrift 2007:10).

Infrastructures fit this category because although critical to the provision of essential services seldom do they get the attention they deserve.[9] Rarely are they considered thrilling sites of imaginative endeavour. The fact that they conventionally take the mode of roads, pipes supplying essential services (such as water or sewerage) or cables for electricity and phone data means they are either hidden deep in domestic architecture or if they are not concealed are rendered invisible by their sheer ubiquity. Star (1999) has done much to advocate the study of infrastructures by pointing out that it is not the infrastructures themselves that are 'boring', but how we tend to look at them.

Because WiFi is commonly known for facilitating the transfer of high-speed data, and is most often associated with the internet, it is largely seen as an ICT infrastructure, or it is not seen at all. The problem in overlooking infrastructures like WiFi, and assuming they are all the same, mundane and boring channels for more exciting content-based applications is that we neglect how they shape and are shaped by society and culture. STS provides a rich framework for the investigation of small and seemingly marginal and mundane things. Michael explains that "'mundane' refers to those technologies whose novelty has worn off; these are technologies that are now fully integrated into, and an unremarkable part of, everyday life' (2000:3). Doors, sewers, seatbelts, velcro and water pumps are just a few STS examples that attend to the idea that seemingly unremarkable artefacts and systems make explicit the familiar and taken for granted ways in which people make sense of and operate in everyday life (Latour 1992; Michael 2000; 2006; Mol 2002). Broadly speaking, these studies hold that a close examination of

DOI: 10.1057/9781137312532.0007

such intertwinings provide valuable interventions in the larger dynamics of socio-technical systems. A distinct approach to complex networks is provided by Actor Network Theory (ANT), which emerged from early STS in recognition of the roles played by non-humans as well as humans in complex networks (Callon 1986; Star 1991; Latour 1990, 1992, 2005). Rather than privileging the role of technology or that of society in the shaping of a new artefact or system, it contends that both human and non-human actors are equally constituted and powerful. In this book, it means taking the weather, backyards, sticky tape and trees as seriously as the actions of makers in the process of understanding WiFi.

Aside from being boring, another problem lies in the practical difficulties of studying infrastructures. Star notes that they are often only visible when something breaks down and even then, they 'tend to be squirreled away in semi-private settings or buried in inaccessible electronic code' (1999:378). WiFi appears to be an extreme case of this, as Mackenzie notes: 'Unlike the dazzle of Hollywood cinema's digital effects, the startling mobility of images in recent computer games, or the efflorescent sociality of mobile phones, WiFi is hardly spectacular in any way, shape or form' (2005b:2). To counter this, one way of studying the complexities of ICT infrastructure is to examine how it is rendered visible and material by the people who make and use it. Haring's (2007) study of male ham radio hobbyists in the United States in the 1950s focuses on a vast body of amateur publications such as club newsletters, technical handbooks and local media journalism. She argues that this 'technical culture' had direct implications in the shaping of social, technical and gendered encounters in the broader radio technology industry. Critically, in the context of this book, Haring's work shows how seemingly boring, invisible and mundane technical infrastructures and cultures are important because they permit 'broader questions of how we think about and think with technology' (162).

Re-imagining WiFi

This book draws on the work of cultural studies and STS scholars who view the internet as firmly embedded in distinctive socio-cultural environments and that to understand it we need to examine the many forms it takes (Miller and Slater 2000; Goggin 2004, 2007; Ito et al. 2005; Goggin and Gregg 2007; Burrell 2009). This marks a distinct shift from early internet studies which focused on connecting somewhere *else*. Caught up in the excitement of a new digital technology many focused on other worlds and identities made possible through the internet (Turkle 1995). In response, studies of everyday interactions in specific places such as internet cafes (Wakeford 1999; Burrell 2009) and national contexts (Miller and Slater 2000; Goggin 2004, 2007) sought to establish the local in the global.

This approach has generated growing interest in alternative versions of the internet, particularly in contexts outside Europe and the United States. Aware that considering national versions of the internet may at first seem 'strange', Goggin points out the imperative 'to attend to its particularities' in order 'to understand where we place ourselves in our society and where we fit in the world' (2004:9). This builds on Miller and Slater's argument that while we may not have benefited from early accounts of the internet as placeless, we 'can gain hugely' in comparisons between places (2000:10). Miller and Slater approached the internet in Trinidad as an everyday social accomplishment made up of material arrangements, relationships and a local understanding. What they were studying was not 'people's use of 'the internet' but rather how they assembled various technical possibilities that added up to *their* internet' (14; emphasis in original). In this context the internet was not a monolithic 'cyberspace' or a virtual experience separated from physical place. It was deeply embedded in Trinidadian culture – it was simply part of 'being Trini' (2000:1). Considering the internet in this way meant it was possible to approach it using ethnographic methods. Drawing on their work, Meikle has argued that 'in this sense, perhaps, the internet has not yet been invented, but rather is always being invented and reinvented in each new context and situation' (2004:75).

Drawing on these studies, I argue that WiFi is not the same everywhere. Yet, because it is mostly known for providing access to the internet, it is often overlooked. Goggin and Gregg argue that 'while there has been a great deal of academic, community, government and industry work on digital technologies in Australia, and much important critical and scholarly work in particular, in our minds wireless technology and cultures have not been given the sustained attention they deserve' (2007:43). It is an idea, however, that is slowly gathering currency; as Goggin writes, '[T]here is something exciting in the prospect of user-led wireless networks leaping over the limits of current home-based wireless routers and playing a real role in shaping future networks' (2007:127). Just as Miller and

DOI: 10.1057/9781137312532.0007

Slater's detailed study of internet use in Trinidad produced broader cutting-edge insights about *the internet*, a deep understanding of a community WiFi network will reveal meaningful insights about larger issues, anxieties and possibilities of *global wireless networks*. This approach explicitly reinforces the idea that the global starts with the local. In other words, community WiFi makers matter. The rest of the chapter discusses the three key themes in more detail.

Critiquing connectivity

WiFi is largely presented in dominant media, policy and commercial discourses not only as a way of *getting* connected to the internet, but of *always* being connected. Scholars have pointed out the disjuncture between these kinds of depictions of connectivity and reality of everyday use (Goggin and Gregg 2007; Gregg 2007, 2011; Mackenzie 2007; Forlano 2008). Broadly, they argue that an uncritical acceptance of them is misleading and limits the possible use and imaginings of wireless technology. As Forlano writes, it 'ignores the particular local characteristics of communities and the specific practices of users' (2008:1). Furthermore, it overlooks place and 'places matter' (Wakeford 2003; Oudshoorn 2012). Mackenzie has gone so far as to term it 'over-connectedness' (2007:94) and Goggin and Gregg have shown that the desire for constant connection in and of itself is an 'increasingly dangerous form of common sense' and call for researchers to 'challenge the growing consensus that citizens need to 'be connected' to fully participate in and enjoy the benefits of a modern democratic society' (2007:42). A particularly notable example is provided by Gregg (2007, 2011) who examined the impact of mobile technologies, including WiFi, on the labour market, and argued that these 'freedoms' bring about new anxieties, specifically concerning the labour politics of a 'flexible' and, in many cases, 'dispensable' workforce. 'Always on' connectivity in her work highlights an inability to switch off. In response to Green and Harvey's question: 'what *disconnections* are entailed in connecting?' (1999:12), Gregg's work draws attention to a disconnection from disconnection. While this study and others like it critically examine the nature of connectivity, drawing attention to the many important and pervasive ways it shapes and is shaped by sociopolitical, gender and technical relationships, there is still space to further unsettle the certain and stable idea of constant connectivity the way, for

DOI: 10.1057/9781137312532.0007

instance, Wyatt et al. (2002) unsettles the category of 'user' by attending to the 'non-users', 'former users' and 'never users' and Bruns and Jacobs (2006) fragment 'blogs' into 'diary blogging', 'corporate blogging', 'community blogging' and 'research blogging'. There exists an opportunity to contribute to this body of work by teasing apart connectivity itself, to see what other forms of connections and disconnections are possible.

As indicated in the introduction, becoming disconnected did not affect every WiFi group member in the same way. Some did not notice a change in service. Others, like Dan, were disconnected from gaming, file sharing and emailing activities. For Tim and Ron it catalysed a spate of late-night innovative rooftop problem solving involving a constellation of at-hand materials, improvised methods and skills related to location, local fauna and time of day. Members' experiences of the network are concurrently connected, partially connected and disconnected at the same time thereby offering a way to interrogate meanings of connectivity. Air-Stream's WiFi network is clearly not 'always on'. What might a group that designs, builds and fixes its own WiFi network reveal about the nature of, and possibilities for, alternative modes of connectivity?

The visual culture of a new digital technology

Although WiFi makers and ham radio operators both use radio waves to make rich social networks, a key difference lies in how they represent themselves. Ham radio is defined by voice connections, a point highlighted by 'SK' in the user's basic handbook, meaning 'Silent Key' and a euphemism for 'deceased'.[10] In contrast WiFi is intensely visual. The study of WiFi provides an intriguing case in the study of representations of knowledge precisely because it is invisible to the human eye. Axiomatic to commercial and community wireless organisations alike is the presence of a plethora of maps, diagrams, stickers, artefacts, websites, photos and drawings designed to show people where WiFi networks are located and how to use, build and buy them. This is particularly evident in the growth of commercially provided mobile broadband; the green bars on handheld devices providing an indication of the possibility of connection. What makes WiFi particularly interesting is the fact that it is hard to pin down; it resists being contained or held in one place and shifts and changes depending on times of day, weather and location.

DOI: 10.1057/9781137312532.0007

Because community WiFi networks rely on point-to-point connections over long distances, it can become easily disconnected. Interruptions can take many forms; from changing weather conditions and growing trees or local bird life to microwaves and even baby monitors that share the spectrum. As a result, it can seem unpredictable and temperamental, shifting from one place to another.

The value of representational practice and visual knowledge is firmly established in STS, particularly in science, engineering and architecture (Latour 1987, 1990; Lynch and Woolgar 1990; Cartwright 1995; Henderson 1999; Latour and Yaneva 2008; Myers 2008). Here, graphs, models, drawings and images are seen as pivotal in understanding how people construct knowledge, design new technologies, organise action and recruit people. Two of the most crucial properties of a representation or 'inscription' are that they are 'immutable' and 'mobile' (Latour and Woolgar 1979). The fact that they do not degenerate when reproduced and can be widely distributed means they can be used to build up persuasive and powerful arguments. Henderson has argued that visual culture is 'a particular way of seeing the world that is explicitly linked to actual material experience in rendering that world' (1999:9). It is 'the glue that holds communication together' (59). The way representations stick people together determines the nature of certain activities, defining what is and is not acceptable. Broadly speaking, this literature holds that techniques of rendering are interlocked with particular ways of seeing and knowing the world.

STS scholars also pursue less certain entanglements of seeing and knowing and messier representational practices (Mol 2002; Latour and Yaneva 2008; Garforth 2011; Street 2011). Mol's (2002) work on atherosclerosis illustrates a disease that evades neat and defined representation by taking multiple forms, not all of which can be seen or understood by everyone. But this does not result in fragmentation or dissolution of the power of knowledge. It does not mean that the surgeons cannot operate or that people cannot be diagnosed or, for that matter, healed. Mol explains how the disease is made to 'cohere' in an assemblage of visual representations and practices. Similarly, Street's (2011) research in a Papua New Guinean hospital examines multiplicity and ambiguity in patient medical records or what she calls 'artefacts of not-knowing'. Rather than isolating a single diagnosis, she describes how medical staff work to alleviate patient's individual symptoms which maximizes 'multiple pathways to action' (10). This work paves the way for findings in

DOI: 10.1057/9781137312532.0007

this study that attend to the production of knowledge that emerges from what can both be seen *and* not seen.

Examining the way WiFi is made visible and material is critical not only for locating and understanding how it operates, what it shapes and is shaped by, and how it is positioned in larger socio-technical and cultural ecologies but also for the way it 'constricts and constructs the literal ability to see or imagine' (Henderson 1999:26). If representing the presence of WiFi in one place is not a guarantee that it will still be there in an hour, let alone in a week, how do WiFi makers make sense of, communicate and expand their network? What role, if any, does the group's visual culture play in shaping who can and cannot participate?

DiY (and Do-it-Together) technology practices

WiFi makers are engaged in DiY technology practice. In this book, DiY is viewed as a hands-on physical engagement with a diverse set of materials and improvised methods for the purpose of creating or repairing something for not-for-profit use outside traditional technology innovation times and spaces. Traditionally, DiY has been linked to experimental dance culture and environmental activism (McKay 1998); land, housing and transport protests (Searle 1997); rare trades such as stonemasons and tinsmiths (Thomson 2002b); and home improvements (Shove et al. 2007:43). More recently, it has become associated with culture jamming (Jordan 2002), hacker culture (Wark 2004), environmental citizenship (Ellis and Waterton 2005), digital culture remixing (Lessig 2004), ham radio hobbyists (Haring 2007; Dunbar-Hester 2008), music makers (Lingel and Naaman 2012) and maker faires.[11] Although this body of work addresses a wide range of subjects and technological artefacts, they are united by an interest in developing a richer understanding of different communities of knowledge that operate independently of dominant commercial and governmental practice. In these contexts practice is not just what people do but a coordinated series of activities held together by norms and performances (Warde 2005).

The book presents a fresh take on DiY via the localised collective practices of backyard technologists undertaking sophisticated digital work in mundane and everyday sites. Backyards, in this context, are concurrently vital sites for wireless technological innovation and more broadly symbolic of a hands-on, resourceful and collective approach.

DOI: 10.1057/9781137312532.0007

While DiY has conventionally been regarded as an individual pursuit, as indicated by its moniker, Air-Stream members collectively make WiFi. A wireless network cannot operate without multiple distributed points, nor can an antenna be raised without the help and support of a community of people. They have to Do-it-Together (DiT). The book describes a DiT technology culture; an approach that marries a collaborative social engagement with a willingness to tinker predicated on an understanding of WiFi as open, malleable and participatory. If DiT is a distinctive way of imagining and making WiFi, what role do collective technology practices more broadly play in an increasingly networked society?

Summary

Framed by the discussion in this chapter, the book focuses on a distinctly suburban community WiFi culture with its unique constellation of materialities, social encounters, the weather and fauna and flora in order to offer insights into larger socio-technical landscapes. It moves on from the limited notion of ubiquitous or monolithic one-size-fits-all wireless infrastructure, arguing that to understand WiFi requires in-depth engagement with local versions deeply embedded in and shaped by ordinary everyday contexts. Bringing to light a distinct version of WiFi contributes a richly nuanced understanding to the global wireless technology scene, opening up the possibility of future comparative studies. In the next chapter I discuss the methodological approach and emerging epistemological issues.

Notes

1   In October 2013, typical Australian download speed was 13.49 Mbps and upload was 2.63 Mbps (see www.netindex.com).
2   Battersby (2013) writes about a new NBN user in Melbourne with 47 Mbps download and 18 Mpbs upload speeds.
3   Australia is ranked 52nd in the world for value for money / cost per Mbps (see www.netindex.com).
4   WiFi is a moniker for a series of Wireless Local Area Network (WLAN) 802.11 standards developed by the Institute of Electrical and Electronics Engineers (IEEE).

DOI: 10.1057/9781137312532.0007

5   See BBC 2004; 2005a; BBC 2005b for a few examples.

6   WiFi operates on different radio frequencies around the world but they are all publicly open spectrums.

7   For comparison, Haring (2007) was able to study the technical culture of male ham radio hobbyists in the US in the 1950s via a vast body of data sourced from national licensing bodies. While there is a growing body of journal articles on wireless technology cultures and edited collections on a global compendium of engagement with wireless applications (see Goggin and Gregg 2007; Forlano 2008; Foth 2009; van Oost et al. 2009; Söderberg 2011; Foth et al. 2012; Powell 2012), at the time of writing there are no books dedicated to a sustained discussion of community wireless broadband innovation.

8   Quotes appearing throughout the book come from up to four in-depth semi-structured interviews I conducted with each member of the community WiFi network group. More about the methodology can be found in Chapter 3.

9   Improvised repair and tinkering practices also fall into this category of neglect (Graham and Thrift 2007).

10 See http://www.arrl.org/ham-radio-glossary

11 See http://makerfaire.com/

DOI: 10.1057/9781137312532.0007

3
Studying Backyard Technologists

Abstract: *This chapter begins by introducing the group, key characters and locations. Despite the prevalence of suburbia in Australian geography, backyards do not play an important role in constructions of national identity. I frame the study by arguing that mundane ordinary spaces, and attending materials and practices, play an important role in the development and understanding of everyday life and relationships to ICTs. I discuss the use of ethnographic methods and the challenges of dealing with constantly unfolding field sites and highlight some of the epistemological, methodological and practical issues that shaped fieldwork such as being documented as I documented others.*

Jungnickel, Katrina. *DiY WiFi: Re-imagining Connectivity.* Basingstoke: Palgrave Macmillan, 2014.
DOI: 10.1057/9781137312532.0008.

Background to the research

TIM: The initial driving force apart from the general playing with computers and stuff was the fact that Adelaide in particular was behind the eight ball with regards to broadband technology. ADSL had only just been introduced. Cable internet, we slowly found out, was never really going to be rolled out in Adelaide. Whereas the eastern states were getting it really heavily rolled out by Telstra. So we thought we wanted to have interconnectivity between the group of us and we all sort of geographically lived in a similar area and after a bit of researching we started to plot out how to do it.

This is how Air-Stream began in 2001. Tim was a keen Local Area Network (LAN) gamer who regularly attended and occasionally hosted LAN events, whereby hundreds of enthusiasts would congregate in a school or community hall to connect computers via tangles of ethernet cables, eat pizza and play games. At that time, Australia had very limited internet diffusion. Less than 20 per cent of households had internet access at home and only a fifth of these had a permanent connection; the rest were dial up (ABS 2001). Those who were connected paid high prices for low speeds and capped download limits. Australia lagged behind many other nations. For comparison, speeds in the UK averaged 13 Mbps while in Australia they were closer to 1 Mbps (SMH 2006). As noted by Tim, Adelaidians were also aware of inequalities in services across the country. Frustrated by the irregular opportunities for group gaming coupled with limited internet access, costly services and the bi-directional up/downloads, Tim and friends turned to WiFi to explore another way of getting connected.

More than a decade later, Air-Stream not only still exists but also continues to grow in parallel with cheaper and easier access to commercial internet provisioning in Australia. This makes it an interesting subject for investigation. Between 1998 and 2007, household internet access had more than quadrupled from 16 per cent to 67 per cent (ABS 2008). In 2007, there were 70 members, 23 major 'backbone' antennas with a further 30 smaller 'client' ones. In 2013, the group had 100 members and 46 backbone antennas. Morris (2004) has argued that LAN events developed in Australia due to a lack of affordable high-speed internet access, and it is possible to see a similar drive behind Tim's initial forays into WiFi. However this does not account for why the group has endured. Where similar groups around the world have diminished, Air-Stream's

DOI: 10.1057/9781137312532.0008

continued presence signals that it offers something different to the dominant commercial model of the internet.

Central to the core premise of this book is the idea that WiFi is not the same everywhere and community WiFi networks shape and are shaped by people, place, politics and materials. One of Air-Stream's unique features is that it is not built for the purpose of sharing *the* internet. While many community wireless groups around the world use WiFi to provide free or low-cost access *to* the internet, Air-Stream are essentially making their own version *of* the internet, hence the description, 'Ournet, not the internet'.

Although a grey legal space exists in Australia for people to share the internet access across a WiFi connection for not-for-profit use, it is not the focus of the group. Instead, the Air-Stream network uses WiFi to build wireless connections between distributed antennas across the city for the purpose of facilitating the sharing of information resources such as websites, e-mail, audio, video, multi-player gaming and other forms of internet protocol (IP) communications. Rather than simply adding content *to the internet*, Air-Stream members are building the very infrastructure *of their internet*, shaping it according to personal interests and needs, the nuances of the landscape and availability of resources.

Ron: We are a little bit different to what's happening in other countries... [where the] internet is more in the culture and is available and accessible and a little bit cheaper. In the States it is everywhere and you often have hotspots where you have free access to the internet, so those wireless communities are more about propagating internet access. But in Adelaide that's not the case. There aren't open networks that you can just tap into. So we have pretty much gone down a completely different path in that we are creating a network pretty much similar to the architectures of the internet itself.

My decision to study Air-Stream was informed by this unique proposition combined with the group's regularly updated and comprehensive website that featured maps, photos, diagrams, forum, 'how to' guides and technical documentation as well as links to other local and international community groups. In 2006, Air-Stream was one of the most active WiFi groups online. Internet researchers have drawn attention to the challenges of studying web-based groups (Kotamraju 1999) and if a WiFi community website is no longer updated or does not exist, it is a clear indication that the group behind it is no longer operating. Guided by STS literature that recognises representations as an important locus of knowledge, the dynamic website identified Air-Stream as a

DOI: 10.1057/9781137312532.0008

group vigorously engaged in the making of WiFi. The group's site also indicated that not all WiFi activity took place online. Makers met face-to-face at annual 'Open Days' where they demonstrated the network to new people, held antenna 'shoot outs' to test the efficacy of home-made devices, ran monthly meetings at the public school as well as regular antenna installation and maintenance sessions. Several founding members remained involved and the group welcomed new people. I made contact in January 2006 and visited the city in March to conduct initial interviews and observations. I continued online research from London until I moved to live full time in the city from June 2006 to March 2007. Over the following two years I returned regularly for up to a month per visit and maintained observations, participation and correspondence with the group via their website, forum and email during this period. I continued to audit the group's activities, visiting and conducting further interviews, the most recent being March 2013.

Making local connections

Air-Stream inhabits the suburbs of Adelaide. With a population of 1.26 million, Adelaide is one of Australia's smaller capital cities[1] (ABS 2011). The city differs from other capital cities with regard to its size, history and urban plan. Over three-quarters of the state's population lives in Adelaide, making it more suburban than any other Australian capital city. Adelaide was also Australia's first planned city. Founded in 1836 and designed by Colonial Light, it retains older style architecture, a thick green belt of public parklands, wide leafy streets set out on an organised grid and sprawling suburbia of predominantly single storey local stone constructions wrapped around a small dense city centre. Another unique feature of Adelaide relates to its backyards. Hall's (2007) comparative study of old and new suburbs signals that the Australian backyard is shrinking everywhere *except* in Adelaide. According to Hall, an average house, which once occupied 30 per cent of an average suburban block, can now take up to 71 per cent. The impact of this expanding footprint is social, visual, environmental and cultural. A larger house, he argues, fosters an indoor mentality. People are spending more time inside, use more energy to cool their houses and consume more resources than people with outdoor space. An exception to this otherwise nationwide shift is found in Adelaide. Hall's (2007) study on the distinctiveness of

DOI: 10.1057/9781137312532.0008

Adelaide's backyards, together with the fact that suburbia comprises such an essential part of everyday life here, supports my choice of field site for the study of backyard technologists.

In addition to Tim and Ron I spent most of my time in the field with 14 people involved in the group, 12 of whom were members spanning an age range of 18 to 50, at school, university, working in part-time unrelated jobs, at local ISPs, volunteer IT positions or in large telecommunication organisations. Together they represented a cross section of the group in terms of length of membership, age, gender, experience and life-stage. These people welcomed me into their homes, invited me to installations and maintenance events, described in detail their hand-made antennas, taught me how to search for wireless signals and regularly explained complex WiFi jargon.

Motivations for being involved were as broad as the age range. All worked, studied or were interested in IT-related fields: keen gamers, open source software community members, committed DiYers, local ham radio operators or simply interested friends. Being volunteer based meant individuals who invested ideas, time, skills, experience, materials, locations and money shaped the network. As indicated in the introduction, technology cultures change in line with shifts in availability of equipment, infrastructure and people involved. This applies as much to community groups as to ISPs in the broader technology landscape. This book does not set out to document in detail what happened prior to 2006 or after 2009, except to highlight distinct shifts related to the different phases of DiY in relation to networks of social, technical and cultural actors. In addition to observing and participating in the group, I conducted up to four in-depth semi-structured interviews with each member. As per conventional research guidelines, I have changed all personal names. The name of the group however, with permission of members, remains unchanged, as the suburban Australia location is critical to the central argument.

Locating the study in the 'vast and unexplored suburban tundra'

Despite the central positioning of the desert, the ocean, the bush and vast open spaces, 'going bush' or 'walkabout' in cultural imaginaries,[2] the practical reality of everyday living in Australia is significantly more

DOI: 10.1057/9781137312532.0008

coastal and residential. To better grasp the idea of suburbia in Australian culture and the role it plays in shaping technological infrastructure it is important to understand a little about Australia's population distribution. Twenty-one million people live on a continent that covers almost eight million square kilometres. This means there are not quite three people for every square kilometre in Australia. Despite this vast landscape, nearly two-thirds of Australia's population lives in capital cities located near the coast (ABS 2011). The fact that the majority of Australians inhabit a tiny coastal edge of a massive land mass means that 'whether we like it or not, the suburbs and suburban life reflect and reproduce stories of being Australian' (Elder 2007:298). As much as Australian's imagine themselves otherwise, suburban living is a quintessential characteristic of being Australian.

The Australian reality might be suburban but for many it is, as Weber writes, 'both their greatest aspiration and their worst nightmare' (1992:24). Similarly, Elder points out how 'the suburbs have been vilified, lampooned, eulogised and idealised' (2007:298). This ambivalent relationship is in part fed by the *Great Australian Dream*, which mythologizes owning a piece of Australia which is most often packaged as a house with a small yard or increasingly a compact apartment. The fact that Australian culture draws more on the dramatic and untamed image of the outback and the people who live there and much less on the unremarkable everyday aspects of suburban existence suggests little is actually known about 'real' Australia.

> By the time of Australian Federation in 1901, almost 70 per cent of Sydney's population were living in the suburbs; a statistic that suggests that despite prevalent and enduring images of the bushman and the ocker, the 'real' Australia was, and still is, more likely to be located in what Barry Humphries has described as Australia's 'vast and unexplored suburban tundra'. (Turnball 2008:15)

Moreover, it is evident that what Australians do know they tend to treat with contempt and cynicism. Writing about one of Barry Humphries' enduring characters Dame Edna Everage, otherwise known as the average suburban housewife, and Kath and Kim, two more recent comics, Turnball argues that suburbia remains a resource for comedy because it draws on a 'long tradition of anti-suburbanism' (2008:15). This is all the more incongruous considering, as MacKay points out, it is not the case that we have gradually become suburban, but that 'we've never been a rural people' (cited in Marks 2004). It is therefore all the more unusual that 'we do not celebrate the virtues of suburban living' (McKay 2008).

Suburbia is easy to overlook because it is right in front of our eyes, functioning as a vital component of everyday life. According to STS, the fact that it is boring and vastly underestimated signals its potential for dispensing vital knowledge about the practices of everyday life (Latour 1992; Star 1999; Michael 2000, 2006). Therefore, it makes sense to study it from one of its central points – the backyard. Identifying these spaces and places is important because Australians' attitude to technology remains influenced by a legacy of bush history even though very few live in the bush. The narration of taking things apart, of tinkering, customisation and technological imaginings of a people who 'have a go' is an enduring trope that shapes understandings and use of new applications and devices. Given the reality of suburban living, the book explores the idea that this enduring bush legacy is translated into the backyard.

Backyard technology cultures

> We continue to embrace the rural mythology, which is really powerful in terms of Australian identity. In a sense, the yard has kept us in touch with the land. (McKay 2008)

Despite the prevalence of suburbia in Australian cultural geography, backyards do not play an important role in sociological, STS or cultural theory. Although there is a wealth of material for talking about suburbia, there is far less about backyards. Most technology innovation stories attend to 'frontyard' technologists in authoritative fields of science and engineering or what Star has called 'heroes, big men, important organisations or major projects' (1991:12). Smaller, less triumphal and more mundane backyard technologists, who build artefacts and systems for non-commercial use, tend to slip by unnoticed. As Thomson, a social historian and director of the Institute of Backyard Studies (IBYS), notes: 'To some, a "backyard operation" is synonymous with dodgy, low quality, illicit and generally dubious business' (2007:2). Yet, mundane qualities are valuable and technological innovation also takes place outside conventional institutional frameworks (Mol 2002; Michael 2006).

Thomson (1999; 2002, 2002a, 2006, 2007) has written at length, and in a scholarly vein, about men, innovation and Australian shed culture, producing documentary record of this unique socio-cultural intersection.

DOI: 10.1057/9781137312532.0008

He steps inside sheds and into the lives of men who fill them with a compendium of stuff. Sheds take many shapes and sizes and shelter a diverse array of tools, materials, machinery and random accoutrement. They are not defined by the specifics of what they hold or even what they are used for but rather for what they enable. Thomson posits that sheds are closely linked to a DiY ethos of practicality and ingenuity: 'In the shed, the rules are different. Here, chaos is allowed to reign, asserting its creative force in wayward contrast to the suburban order all around' (2002:2–3).

Bell and Dourish (2006) also examine sheds in suburban culture. They argue that sheds offer critical vantage points into ICT relationships – providing a privileged viewpoint on domestic practices and gendered relationships around technology, enabling new ways to think about socio-technical relationships. Sheds are sites of encounters between new technologies and existing domestic ecologies, at different points in their user trajectory. Artefacts move from the house to the shed when they are broken, unsafe or have lost their initial use. In the shed they are fixed and returned to the house, or dismantled, given a new lease of life doing something new or left in pieces, to gather dirt and dust for the right moment. 'In some ways', Bell and Dourish write, 'one might also regard the shed as a very real staging point for technologies coming into or out of the home – it is a place for not yet domesticated technologies or for those that must forever remain feral and dangerous' (2006:375). What is interesting here is how the shed operates as a lens for looking anew at stuff, that may or may not work as intended, and for imagining new application. '[F]or as much as sheds function as sites of particular activities, they are also a cultural form; an imaginary realm within the larger domestic expanse' (ibid).

This work points to the importance of mundane, ordinary and other-wise overlooked architectures of domestic life. Although deeply embedded in suburbia, sheds and backyards carve out creative space, harbour stuff to materially think through ideas and provide time to dwell. They are simultaneously critical sites for wireless work and emblematic of a hands-on, resourceful and collective approach. This study takes backyards seriously on both accounts. If sheds function on the edge of the domestic, backyard technologists operate on the fringes of large scale centres of innovation. Just as the 'edgefulness of sheds might allow a very different understanding of the "home"' (Bell and Dourish 2006:376), it is feasible to consider the backyard technologist as enabling a new

DOI: 10.1057/9781137312532.0008

understanding of wireless technology. The research also recognises sheds as spaces where masculinity is produced as a counterpoint to the conventional narrative of the home as a feminised space. The 'barbie' or barbeque, which features strongly throughout the book, is renowned for reflecting and producing forms of masculinity. Because these spaces and practices are prominent, gender is an underlying thread in this study of WiFi.

Ways of seeing and knowing

Practically, to understand how WiFi was made I followed members' activities across the city, various technological devices and the internet. In addition to backyards, field sites included residential houses, rooftops, garages and sheds, community centres, public schools, members' workplaces, cafes, libraries, websites, blogs and online forums. I spent time on the phone, texting, instant messaging, emailing, posting on websites, talking, taking photos, attending meetings, volunteering at open events, helping to raise antennas, build new devices, learning to fix others and sometimes just standing around holding things. I took hundreds of photos and made detailed notes of my experiences.

In addition to the vast publicly available website, I gained access to the member's section featuring personal blogs, forum, issue tracker, maps, technical schema and an email account through which I learnt about upcoming activities. I attended bi-monthly committee meetings as well as the monthly members' meetings. It is important to note, and is discussed in Chapter 4, that membership was not a pre-requisite for viewing or contributing to the general website or attending the many group events. However, for me, it provided access to even more of the group's visual and cultural practices. In addition to meetings and public Open Day events, I volunteered at antenna-raising installations, maintenance and tinkering sessions, attended 'Tech Nests' in member's backyards and 'stumbled' for digital noise. I also attempted to get connected to the network from my house, and although unsuccessful it provided a means for understanding the process through which new antennas are linked into the infrastructure and the network expands across the city. Importantly, and will be discussed in detail in Chapter 6, it illustrates how being technically disconnected did not prevent me from participating in and contributing to the group.

DOI: 10.1057/9781137312532.0008

Because WiFi is physical, material and social, I used ethnographic methods. Ethnography involves engaging in everyday activities, paying detailed attention to interactions and developing relationships with people in key settings over a period of time. It combines a range of qualitative methods as a means of gaining an in-depth understanding of a culture from the point of view of its participants (Emerson et al. 2011; Fetterman 2010; Hammersley and Atkinson 2007). Importantly, for a study of mundane and ordinary spaces in the form of suburban backyards and a technology that is commonly considered an infrastructure and invisible, ethnographic methods offer a way to develop a rich understanding from multiple perspectives – observation, participation, interviews, visual analysis and personal reflection. Using a grounded theory approach enabled me to constantly check my emerging ideas against further experiences with WiFi makers (Glasner and Strauss 1967).

During my fieldwork I lived on the north-west fringe of the city centre in a stone-and-brick dwelling with front and backyards, typical of the suburban architecture of the area, in easy cycle commute to the majority of meeting points, members' houses and other antenna sites.[3] From the outset, I joined the group as an overt researcher and volunteer member. This meant members were aware of my research project when they accepted me in the community. Lofland et al. note one of the initial difficulties in observing and participating is 'getting in' to a group or site of study, which they define as 'gaining the acceptance of the people being studied' (2004:20). Inside positions provide ethnographers with intimate access to integral practices and associated activities potentially unobtainable to a distant outsider. Unlike other ethnographies whereby the researcher has to remind respondents of their status (Hine 2000; Silverman 2004), there was little ongoing ambiguity about my presence as a researcher for a number of reasons. I was the youngest woman at 33 years of age, my knowledge of WiFi was limited compared to other members and I was new to Adelaide. The former is relevant because out of 70 members in the group, 67 were men and three were women, including me. Also, despite being an Australian national, I had an accent from living in London for a decade. I had two further potentially alienating characteristics; I was a cyclist (everyone else drove or negotiated car rides to events) and, given the group's proclivity for barbeques, a vegetarian. Yet, these factors did not make me an outsider. It worked in my favour as members actively sought to describe social, technical and cultural features that they thought I would not understand, bringing to

DOI: 10.1057/9781137312532.0008

light the rich DiT culture of the group. Often multi-narrative and contra-dictory in nature, their collective making and communication practices rendered visible a stickiness that held the group together.

Mess and the potential for other kinds of description

One of the key features and challenges of ethnography about digital tech-nology is the nature of a constant 24-hour field site in which researchers contend with practical, conceptual and methodological challenges (Hine 2000). I did not anticipate this being an issue for my study. Nor did the WiFi makers. If anything, they expressed concern that there would be enough for me to study and wanted to ensure I had things to do between WiFi activities. There was of course plenty to do and on occasion things overlapped. The volunteer nature of the group meant that much hap-pened around and outside normal working hours. For most members this meant evenings, weekends and lunch hours. However, not every-thing happened in one place. Being multi-sited and on and offline meant wireless work often co-existed.

Hess (2002) has argued that 'second generation' STS ethnographies are characterised by a range of field sites accessed via multiple entry and exit points and constant unfolding field sites. The researcher moves around, following ideas, objects, metaphors and people in what Marcus (1998) has termed 'multi-sited ethnography'. In line with the changing nature of contemporary ethnographic study, Marcus' theory articulates a new and better purchase on the complexities of multiple field sites, ideas and things. Researchers maintain a critical eye by stepping in to the point of view of the informant in their situated and then stepping out to achieve an objective view and analytic perspective. This 'back-and-forth movement' (Hess 2002:238), however, is possible only if the researcher *can* step out of the field site. Further to finding Air-Stream's field sites in many places, I found myself engaging with complex overlapping and co-located field sites in the *same* place made possible by their visual culture. One such example was when I found myself sitting at my laptop at my living room table, catching up on field notes from a meeting the night before.

> My notebook is on my right. My camera downloaded photos to my left. Simon popped up in an IM box on my screen. He gave me details to upload photos I took at a recent installation. I opened up another window on my

screen to start attaching images. I was on a slow internet connection so they took a while to load and I continued to type notes. Simon and I chatted and he sent me photos he took of the new antenna at a water tower. Talking about recent scavenging missions he has been on he also sent me links to the council rubbish website. As I wait for his photos to download and for mine to upload, I flick to the web address he has sent and then onto the group website. While I'm there I read the most recent posts to the forum and see new 360-degree photos of the view from the top of the building. I see mention of another meeting so I log in and open up my Air-Stream email to see whether there is any more information. I mention it to Simon and we talk a little about it. I see his photos that have finally downloaded. I make notes about our conversation on another word document. My mobile phone beeps. As I reach for it I am reminded that I'm still sunburned from a recent installation event and I have slightly sore arm muscles from climbing up and down long ladders and winding buckets of hardware up eight concrete floors. I joke about this to Simon and he sends me a video he took of me on the day. I read the text message and see that is from Peter responding to an earlier question I asked about an upcoming event. I text back as I unclip the camera and take the battery out ready for recharging.

This experience provides insight into the nature of the study, how the boundaries of field sites were not firm and I was rarely out of the field. Confronted with multiple overlapping field sites and being documented myself as I documented others was at times daunting. Axiomatic to ethnographic literature is the expectation of becoming 'overwhelmed', 'unnerved' and 'daunted' in the field (Hammersley and Atkinson 2007; Fetterman 2010). Although an essential part of getting immersed in a new field, the researcher is expected to take control, and organize this messiness into a linear sociological argument. Mess in certain circumstances, however, can be seen as generative of new ideas with many writers deliberately disrupting of the notion of field as fixed, methods as static and the respondent as passive in contrast to the active researcher (Whatmore 2003; Law 2004; Lury and Wakeford 2013). Law, for instance, has argued that 'if we want to think about messes of reality at all then we're going to have to teach ourselves to think, to practice, to relate, and to know in new ways' (2004:2). More recently, Lury and Wakeford (2012) have written about methods that cannot be separated from the research problem at hand. They are intimately entwined in the process, shaped by and shaping of key issues. 'Inventive methods are ways to introduce answerability into a problem ... if methods are to

DOI: 10.1057/9781137312532.0008

be inventive, they should not leave that problem untouched' (Lury and Wakeford, 2012:3). Mess loomed large during my research. It was in the field, methods, representational practice and in complex socio-material interactions such as how Simon and I were co-making meaning about the network via multi-dimensional representations. Sharing insights, documentation and links as well as what worked well and less well are all part of being in the network and provides further evidence of the group's DiT culture.

The potential and pitfalls of a shared visual culture

In addition to the shared public website and network, many WiFi makers blogged and were active on forums. This meant digital cameras and note-books were regular sights at meetings. Laptops and video cameras were also common. As a result there was little concern about my documentation practices as they fitted with the group's visual culture. WiFi makers were used to communicating with each other through images, text, maps and diagrams and I found a number of ways through which I both participated in their visual culture and correspondingly dealt with the mass and mess of ethnographic data. In addition to maintaining a research blog, I pro-duced collaged photos and drew on the groups' proclivity for barbeques and open source practices to host an exhibition to elicit responses to my emerging ideas. These practices align with what Back et al. (2008) call new forms of sociological representation that resist 'flattening the texture of social experience'. They are also examples of inventive methods, which Lury and Wakeford describe as the 'means by which the social world is not only investigated, but may also be engaged' (2012:6).

Inspired by British artist David Hockney's 'joiners', my collages emerged in response to feeling that single photos restricted the bounda-ries of events while more images promised texture and busyness in which to explore a range of activity, interactions and actors. They encompass multiple images roughly pieced together in an attempt to capture the object or activity in its larger dynamic and messy socio-material and spatial ecology. Multi-temporal and fragmented, they go some way to reflect the dynamic collective spirit of WiFi activity. I also often included a part of myself (shadow, foot etc.) in the images thereby contributing myself to the mess, which further enacted the DiT culture of the group. In a similar vein, the homemade exhibition was located in a suburban yard around a barbeque. It featured photos, blog posts and field notes using at-hand materials such as clothes pegs, lengths of electricity cable and house nails. WiFi makers literally entered into my argument and took away fragments that interested them (Jungnickel 2006, 2009, 2010). 'Modding' [modifying] my ideas to fit with the space and my audience, in its collaborative, messy, overlapping and co-located manner, is much in keeping with arguments put forth in the following chapters.

This insight, however, took time to emerge. Early in the research several members recognised my interest in documentation and I was asked to take photos at a group event. I was initially hesitant. How would I gain an understanding of the visual culture of WiFi makers, if I were the photographer? Yet, as I discuss in detail in Chapter 4, roles in the group were not rigorously defined. Many people took photos and blogged about events. Much like making WiFi, they documented it together. Taking on the role afforded a way not only to participate but also to contribute to their everyday practice, which in turn provided a lens into how members co-produced, understood and communicated ideas. Members, including resident researchers, could not simply observe and participate. They also had to contribute. Had I rejected the invitation out of fear of contaminating the field site, I would have missed out on a critical insight into the dialogic nature of group's DiT culture.

Another dialogic experience emerged in the way WiFi makers documented my activities in the group as I did theirs. I initially attempted to render anonymous the name of the group as well as respondents. It was soon impossible to do the former as WiFi makers referenced and linked to me on their blogs and websites. In response, I began to document the way they documented me and noted how descriptions changed over time. This tactic mirrored how members familiarised themselves with potentially interruptive elements in the network (such as trees, birds, bugs etc). Rather than distancing them, they diffused potential threats posed from foreign actors by enfolding them into the network.[4] Empirical data in the following chapters brings to light how the group incorporated a diverse range of heterogeneous human and non-human actors into their network as a means of dealing with uncertainty, which in turn serves to make it strong and resilient. They adapt, tinker and hack roles and ideas much like they do technology; making things fit together in ways they might not have been initially designed to go together. In attaching me in various ways to their DiT culture, they connected me to the group; even as I have noted, I never achieved a technical connection to their WiFi network.

DOI: 10.1057/9781137312532.0008

Summary

This chapter has outlined some of the practical, theoretical and methodological challenges of studying a group of community WiFi makers. I discussed the presence of multi-sited objects of enquiry and fields of study that constantly unfolded in multi-dimensional and temporal forms and how I adapted my ethnographic engagement to follow new paths as they appeared, tracing overlaps and intersections. I described how approaching the study not only through participation but also contribution shaped the production and transmission of my sociological knowledge. I also drew attention to the location of the study, highlighting the important yet marginalised view of suburbia as a site of innovation and argued that backyard technologists offer a unique view on relationships with and around ICTs. As Miller and Slater argue, 'if you want to get to the internet, don't start from there' (2000:5); similarly to *get to* WiFi I start with the nuances and textures of Australian backyards, barbeques, birdlife, bugs, the weather, maps and makers. The following chapters tell stories about how individuals collectively make 'Ournet not the internet'.

Notes

1 In June 2011, the population of South Australia was 1.64 million.
2 The 2007 Australian Tourism Commission campaign featured remote and isolated places of extreme beauty. In 2008, Baz Lurhman's version featured stressed, unhappy residents of densely populated cities reinvigorated after going 'walkabout' in remote Australia. The 2000 Olympics' Opening and Closing ceremonies represented Australian history through images of bush, desert and beach.
3 See Jungnickel (2013) for more discussion on how cycle commuting shaped a study of backyard technologists.
4 Online descriptions of me by WiFi makers changed over time. I was initially described as 'A PhD sociology student'. After a month I became 'Our resident sociologist'. When I left the following year I was 'Our resident bicycle riding PhD student'. These shifts in familiarity signal a deepening involvement in the group and means through which they sought to incorporate me into their visual technological culture, as a method of enrolment and acceptance.

DOI: 10.1057/9781137312532.0008

4

The 'Barbie' and WiFi

Abstract: *This chapter is situated at a WiFi meeting. I describe the nature of membership and explain different forms of network connection. Foregrounding the multidimensional, co-located and occasionally contradictory nature of WiFi representations, I discuss how they connect people together, aid recruitment and teach members about new applications. I argue that the resilience and responsiveness of the seemingly scattergun visual culture is well suited to the idiosyncrasies of WiFi makers and their disparate ideas and approaches. I also introduce and explain the role of the barbeque or 'barbie' in the making of WiFi, arguing that it operates as a critical means of contending with the complexities of the technology by domesticating public spaces and cementing social ties.*

Jungnickel, Katrina. *DiY WiFi: Re-imagining Connectivity*. Basingstoke: Palgrave Macmillan, 2014.
DOI: 10.1057/9781137312532.0009.

Stabilising unstable connections

A man stands in the middle of a tarred quadrangle in front of a steel rectangular barbeque blackened by use and packed with neat rows of thin non-descript pink 'snags' (sausages). With a pair of tongs he makes a small gap on the hot plate and shakes out a tangle of onion slices from a plastic container that jump and sizzle in the hot oil. He returns the container to one of the two school tables that have been carried out from a nearby classroom. Three bags of white sliced bread, a plastic bottle of tomato sauce and paper serviettes are arranged across the tables. Eighteen people cluster in groups of three and four nearby. Some hover around open laptops, others peer into boxes of cables and antenna equipment. Some just chat. Rolling back on his thongs the man tries to dodge the spitting fat from the sizzling meat. He fans an edge of his loose cotton shirt in an attempt to cool himself. Summer in Adelaide can be fierce with temperatures hovering in the mid-forties for weeks at a time. Even at seven o'clock at night a thick residual heat baked into the black tarmac underfoot rises up around our legs.

Barbeques or 'barbies' as they are known were often held in summer months before monthly Air-Stream meetings in a primary school on the fringe of the city parklands (Figure 4.1). Being located in a public location with free car parking, on a Wednesday evening and open to non-members is all part of the group's strategy to represent themselves as open and inclusive to a broad interested community. Access to the school was negotiated by an 'old member' of the group who happened to be the school's IT consultant. Despite no longer being involved in the group, the agreement was upheld. Members had access to several single storey stone buildings that edged the tarred quadrangle, such as a classroom equipped with chairs, movable tables and whiteboard, library (when meetings were large, such as the Annual General Meeting), staff kitchen, toilet block and carpark. They were also able to leave equipment in secure storage areas and the rooftop of one of the teaching rooms provided a site for a WiFi antenna.

I start to introduce myself to the cook. Although I have attended meetings for six months, I have not seen him before. Yet, from experience, this is not unusual. People tend to drop in and out of the group due to work, school or family commitments. Out of 70 members, between 15 and 30 regularly attend the monthly meetings. The man interrupts my brief introduction. He says he knows me. He has read about me on the

DOI: 10.1057/9781137312532.0009

FIGURE 4.1   *A WiFi 'barbie' in the quadrangle of the local primary school*

group's website. Jason, 25, I learn, is one of the 'original members' who initially set up the network with Tim. He tells me he has not been very active in the group recently due to a new job in his dad's winery coupled with a flat move. He had to take his antenna 'down' and although he has contacted his new landlord and hopes to reinstall it on the roof of a communal block of flats, going by a few 'site surveys' he's done, it 'doesn't look good'.

Jason is currently disconnected from the network. Disruption in the form of a new flat and job initially brought his antenna 'down'. A 'site survey', otherwise known as 'stumbling', which generates information about the strength and direction of nearby wireless signals to assist in getting an antenna 'up', has been disappointing at the new location. He cannot 'see' the network in order to join it. Because WiFi relies on point-to-point connections it can become easily disconnected when line-of-sight is broken, which is a particularly relevant issue for community wireless networks that operate across long distances. Even if an antenna works in one location, it may not work well, if at all, in another. Yet, despite this setback, Jason does not seem anxious or detached from the group. He knows what is going on; he is at the meeting, cooking the barbeque and knows about the newest member. This is because, as he explained,

DOI: 10.1057/9781137312532.0009

he regularly reads the group's website. It is published on the network for those who are connected and made publicly available on the internet for those who are not. Like me, Jason found out about the barbeque via the agenda Ron posted on the website a few days in advance of the meeting. Ron also published stories about my involvement with the group, complete with pictures and links to my research blog. Jason may not be part of the network, but he is clearly still part of the group.

More people arrive in the quadrangle and Jason and I are drawn into conversation with Kerry, Jan, Julie and Kurt who stand in a semi-circle around the barbeque. Knowing that I have been living and studying in the UK (from the website and previous conversations), Kerry, 35, talks about his and Jan's experiences with London WiFi groups. Stretching his arms out in front of him and bending one leg behind he shows the group how he used to 'rig up' networks across the small alleyways in South London. He leans forward to physically explain the precarious nature and comedic potential of a large man hanging out of a small window at a great height. Everyone laughs. Like Jason, Kerry and Jan are also not connected to the network. The difference lies in the fact that they have never been connected. Despite trying for the past year, they cannot 'see' any of the local Air-Stream signals from the roof of their rented house in the south of the city, due to trees and buildings that block the signal. This however has not stopped Kerry from joining Air-Stream as an 'official' member, who is someone who has paid the annual AUS$50 fee. He regularly posts pictures and stories to the group website of his attempts to get connected and is often called upon to 'demo' (demonstrate) complex technical ideas at meetings, which draws upon his work as an IT manager at the local university, his WiFi experience in London and his warm sociability. Kerry is also often the first to volunteer and share his experience. His contributions were recognised at the 2007 AGM when he was voted onto the smaller eight-member committee that meets more regularly to help steer the group.

Jan, 40, is a self-confessed 'tourist', which is what she calls people who attend meetings without officially joining. She works as a freelance graphic designer and volunteer for local OSS events. Well read about current debates on digital politics and internet freedoms, Jan believes that community WiFi networks 'engineer alternatives' to commercial telecommunications systems. She regularly promotes the group's activities on her personal website in what she calls 'Jan rants' and also on related listserves and noticeboards. Both Kerry and Jan are involved in local IT

DOI: 10.1057/9781137312532.0009

recycling schemes and regularly introduce me online and at meetings to people 'I need to meet'. Even though Jan and Kerry have never been connected to the network, it has not stopped them from being involved in the group, promoting it to others or inviting people to meetings.

Tonight, Jan and Kerry have brought Kurt, a friend and colleague from their OSS activities, to the barbeque. Jan calls Kurt a 'newbie' and he smiles when she does this because they both know even at 21 years of age and new to Air-Stream, he is not new to IT. He has been working for IT charities, technology recycling and adult education organisations throughout high school and more so now he is finished. Tonight he is wearing an OSS T-shirt marked with red dirt which he explains has come from a computer he dismantled that was donated from a company in the desert. He is interested in WiFi and in joining the group even though he knows he 'lives on the wrong side of the hills' and 'off the map' but still hopes to 'find a way'.

The Hills, so called by locals, pose obstacles in the wireless network because they are in fact a mountainous range filled with tall eucalyptus gum trees. Guided by the group's Network Node Map that visually represents the coverage of the network across the city, Kurt is aware it is unlikely that he will be able to connect to the network from where he lives in the shed in the yard of his parents' house. However, this has not stopped him attending the meeting. His 'hope' to 'find a way' suggests his desire to meet experienced people who in turn will help him get connected.

The other person in the group is Julie, 50, a fulltime IT technician at a large commercial telecommunications organisation and one of the most experienced people in the group. She is the owner of a seven-metre steel tower embedded in two square metres of concrete next to an in-ground pool in the sprawling backyard of a house she owns and shares with her teenage daughter in a suburb west of the city. Julie has been a keen ham radio hobbyist since she was 12 years old and an Air-Stream 'committee member' for more than two years. As a result, her tower holds a variety of radio and WiFi equipment. She is one of the most established 'official' members with a sophisticated 'setup', a term used by members to describe the technical assemblage of an antenna, even though she occasionally has to climb up and 'jimmy' (adjust) it to get a better signal.

Membership in, and connection to, the Air-Stream network is far from straightforward. There is no one single type of member or linear process of becoming involved in the group. The barbeque reveals the presence

DOI: 10.1057/9781137312532.0009

of 'old', 'original', 'official', 'committee', 'tourist' and 'newbie' members. The absence of 'non-members' suggests no one is ever excluded from the group and the continued presence of an 'old member' illustrates the absence of a clear exit point as well. There is also no linear trajectory to follow. People can be several types of members at the same time. Jason for instance is an 'official' and an 'original' and for a while he was a 'committee' member too. Further complexity emerges in terms of being 'unconnected', temporarily 'disconnected', 'trying' and 'interested' in getting connected. As Jason and Julie illustrate, even those with sophisticated knowledge and experience become disconnected from the network from time to time. Conversely, Kerry and Jan have never been connected and yet they are actively involved in the group and Kurt, who currently has little chance of connecting, is still at the meeting and expresses interest in joining as an 'official member'. Connection is not a neat or definite achievement or the result of a transaction. It is variable and subject to conditions. Although a primary aim of the group and the central promise of WiFi ('always on') technology, connectivity, like membership in the group, is not a singular concept or it seems even expected in this community.

Clearly, connecting to the network is not easy. It is not simply a case of buying a wireless device and plugging it in. Instead, it appears to involve a significant investment in time, work and interest. This marks a departure from conventional commercial WiFi models. Being a member of a volunteer community wireless group does not predicate being connected to the network. Just as the internet is disaggregated from WiFi, membership to the group is disaggregated from the network. The significance of this point is sharpened if compared to another hobby. Imagine joining an ice skating club even if you had no ice skates or could not get to the rink. Or if you had both but occasionally the rink disappeared and you could not access it. This example goes some way to illustrate what is going on here. Although there is a desire to connect to the network, it is not a requirement of being in the group because as illustrated by Jason and Julie, even achieving a connection to the network does not guarantee staying connected to the network.

Transactions and interactions

As the heat drops the numbers continue to swell and apart from Jan, Julie and I, the rest of the group are men, in two age brackets: 18–25 and

DOI: 10.1057/9781137312532.0009

35–50. Half are dressed in variations of beige trousers and short-sleeved collared shirts, having arrived from office jobs. Others wear shorts or jeans, t-shirts and trainers, with backpacks or computer satchels over one shoulder. Soon there are 25 people in groups of four and five clustering in the quadrangle (Figure 4.2). One cluster of young men surrounds John, 24, who balances his open laptop with one hand and points to images on the screen with the other. John is selling wireless equipment, taking orders at the meeting, and via the website, to make bulk purchases from suppliers in Asia and distributing equipment weeks later at meetings at cost price. Drawing on his reputation for affordable prices, John holds the attention of his audience and sells his product. Images work to represent these objects so he did not have to bring bulky antennas or boxes to the meeting in order to prove they exist. The fact these images did not change from when they were taken to their display at the barbeque give his buyers confidence they represent exactly what it is that he is selling. Without them, John would no doubt have had to work much harder to attract the attention of buyers and achieve sales. He might have, as Latour observed of scientists without their visual aids, 'stuttered, hesitated, and

FIGURE 4.2   *Scorched snags, charred onions and wireless work*

DOI: 10.1057/9781137312532.0009

talked nonsense' or at the very least lacked a certain kind of credibility with his audience (1990:22). Like Latour's scientists, these inscriptions served to 'keep them in their proper place' (22). However, there are other activities taking place at the barbeque that appear contradictory.

Peter, 40, sits at a school desk to the left of the barbeque that is covered in neatly arranged rows of black collared shirts, beanies (woollen hats), and tinnie-holders (beer coolers). All are embroidered in blue with the words 'Air-Stream' near a symbol of radio waves. Peter informs the men clustered around him that the shirts are 'twenty', the beanies are a 'tenner' and 'tinnie-holders are on a meeting special' which includes a can of soft drink for a 'fiver'. Peter is wearing a black shirt that has a different design to items he is selling. He lives in the north of the city and is part of what he and five local members call, The Northern Wireless Order or NWO. They designed their version of the logo and a member used his mum's sewing machine to stitch it onto their shirts.

Like John's photos, this merchandise does not degenerate and can be widely distributed. Yet, in this case, different versions of the same logo keep members not in one 'place', but in several. There are multiple versions of the same representation, some of which appear contradictory. Different logos could be viewed as a division in the network, a splintering of the group into location based portions. Yet this does not happen. They co-exist. The fact that Peter does not even ask if his customers are members or connected to the network means anyone can buy and wear Air-Stream merchandise. Rather than imposing rules and regulations and separating members from non-members, the group's visual strategy serves to strengthen connections between people who want to be involved, rather than identifying those who can from those who cannot join the network.

Recruiting people

Air-Stream barbeques put the network on display and, as a result, complicated aspects of participation are constantly produced and reproduced. Because the agenda on the website invites 'anyone who has a genuine interest in community wireless networks' to the meeting, there are always new faces. New faces, as we have seen with Jason, are not always new people and while some 'newbies' like Kurt attend meetings with members, others arrive on their own. No one however arrives without some knowledge of the network.

DOI: 10.1057/9781137312532.0009

Three men in their early twenties with short-cropped hair, in blue jeans, coloured t-shirts and scuffed trainers with satchels slung over their shoulders wander across the quadrangle to the barbeque. One of them introduces himself as Joel and he tells Kerry, Kurt, Julie, Jan and me that he found out about the meeting on the website but has been interested in WiFi for a long time. He attended the last Open Day. The group occasionally runs events at the school or local community hall where they demo equipment and explain the network to interested people. Joel says he wants to join so he and his friends can play computer games but he is not yet connected because of 'time, trees and apathy'. Kerry nods knowingly. Although he and Jan have invested time and interest they are fully aware of the trees that block their access to the network. He asks where Joel lives. 'Pasadena'. Kerry nods again, saying that he should have no trouble getting connected there. Joel says he knows because he has seen the map, referring to the Network Node Map available on the website and regularly updated to accommodate antennas that go 'up' and 'down'. Kerry points out a young man in the crowd as 'Ben' who is 'the node' at Pasadena whom Joel 'should talk to'. Joel says he has already been in contact with him via the online forum.

Joel's story reveals multiple entry points into the group enabled by a range of publicly available representations of the network. Long before he will get connected, Joel has accessed important knowledge about the group. He has a copy of the current network map, has participated on the forum, attended an Open Day and is now meeting people at the barbeque. Just as there is no clearly identifiable exit, or progression through the group, there is no single entry; there are many. What is interesting is not only the breadth of representations available to people outside the network, but also the lack of a sequential order in which to encounter, understand and collate them.

Joel's experience suggests that each representation of the network that he encountered was designed to operate independently, as well as in relation to one another. These forms of knowledge are not sequential or linear. Although Air-Stream's representations are co-located, occasionally contradictory and do not fit together in neat chains or cascades, nevertheless they work together. Tim explains how this works:

> I guess because what we do is so technical, without some tangible thing for people to visually attach it to it's hard to think about it in your head, especially if it's something that is completely foreign ... there is often a big gap in the understanding of people who come ... and they'll sit through a few

DOI: 10.1057/9781137312532.0009

meetings and just go, 'This is crazy, I can't understand anything' and you never see them again, and that's happening probably more often than I even know about, I'm sure. So [I'm] trying to find a level playing field for everyone and I think by making things visual you are really helping with that … it is so much easier if you've got something there to show them – diagrams and maps and things. Particularly with a map, people can go, 'Oh look I'm there and I can see that'. It pulls them in as well.

What may at first seem to be a messy range of multi-dimensional representations has been deliberately designed to appeal to the heterogeneous nature of an audience located outside and inside the group. Tim knows that the more the representations available to people like Joel, the stronger the attraction and increased possibility they will 'visually attach' to the network and 'pull them' into the group. In view of this, the groups' seemingly scattergun approach appears to attract people in ways that a single, narrow and linear version would not.

Snags and wireless work

Jason does not need to call people over for food because the smell of meat beginning to burn loosens the clusters distributed around the quadrangle only to reform around the barbeque. Hands reach into pockets and coins drop into an old tin wrapped in a hand-written label 'Hi! I'm Mr Cantenna. Please fill me with gold coins to help support Air-Stream'. Clustered conversations do not stop. Curling a piece of sliced bread in one palm, members squirt a line of sauce in the centre and wait, one by one, for Jason to deposit a partially charred snag and some onions on top. People wander slowly away, eating and talking. Clearly, the barbeque is not really about the food. On some occasions less than ten minutes in almost one-and-a-half hours of chatting is dedicated to actually eating (Figure 4.3)

The reason Air-Stream meetings and many other WiFi installation and maintenance events feature a barbeque is only partially for the food and more to do with the social and spatial framework it enables. The barbeque does not simply provide a convenient meal before the start of wireless work; it is a vehicle for wireless work. WiFi is a technology that depends on connecting independent nodes together to form a network. Therefore, it holds that forming relationships with members is as vital as learning about technical specificity. Food creates a domestic ritual. In

DOI: 10.1057/9781137312532.0009

FIGURE 4.3   *The barbie is not simply a convenient meal before the start of wireless work; it is a vehicle for wireless work*

this case it moves the practical into the social. The barbeque domesticates space, providing a platform for the sharing of multi-dimensional demonstrations, story-telling and visual representations of the network; a way of making sense of a technology that is technically invisible. It also transforms the quadrangle from a primary school into a community WiFi group event. Or, moreover, it turns it into a backyard, which is a familiar, non-threatening and creative space for this kind of hands-on technological tinkering. As a result, it does not matter where the barbeque takes place, who cooks or what is on offer. The type of food is not as important as the non-formal, open and inclusive social space that it fosters and platform for the production and sharing of its visual and material culture. The idea that scorched meat is secondary to social connections is reinforced by Thomson, who has written at length about the barbeque and Australian culture:

> An Australian barbeque is an instant excuse for socialising. The barbie is a loose social framework in which many things are possible – an open door for anything from fairly outrageous drunken behaviour to the simple pleasure of eating outdoors in the company of friends. The barbeque has become

DOI: 10.1057/9781137312532.0009

the quintessential Australian social event. This appears to confuse people from overseas, who are expecting some sort of culinary display. They don't realise that a barbeque is more a form of behaviour (in some cases fairly pathological at that) or ritual rather than the cooking of gourmet food outdoors under strict foodie guidelines. It's too bad if it got rained out or you ran out of gas or there was a terrible family fight. You were going to have a barbeque, and that's the main thing. (1999:112)

The barbeque is much more than a burnt snag, a cooking tool or an outdoor appliance. In fact, as Thomson argues, the food might be awful. Its purpose is much larger. It is a ritual so rigorously embedded in the national cultural framework that its original purpose, the very thing it is designed to do (in this case, cooking food) is only one of the many loose boundaries in which it operates. As discussed, being connected or getting connected is not a singular, definite or easy achievement in the group but rather must be constantly produced and reproduced. Given the challenging task that is WiFi networking, the desire for camaraderie and bonding provides guaranteed support and assistance. Social relationships are thoroughly implicated in erecting antennas and maintaining network strength and connectivity. In view of this, the barbeque reflects and produces a crucial social connectivity in the group.

It also accounts for why Kurt attends meetings even though it is technically impossible for him on his own to achieve a connection to the network. Likewise, it explains why Joel had already made contact with Ben, 'the node', nearest him, and also why Kerry did the same for him at the meeting. Connecting to people is critical for finding ways to connect to the network. In fact, the barbeque was such an embedded part of the group, that Ron on occasion used it to illustrate larger concepts such as the Public Park Spectrum:

> We started doing training about the regulations of the ACMA. 'This spectrum is for the public to use. Think of it like a public park.' I remember putting out a number of emails with a picture of a playground and park benches and saying: 'Would you like it if a hamburger truck said I am using this place to sell hamburgers. You can't have your barbeque here anymore?' It had a picture of the barbeque area and park benches and things.

Do men 'stick' to WiFi?

The prominence of barbeques in the Air-Stream calendar provides an unexpected vantage point from which to reflect on the gendered

DOI: 10.1057/9781137312532.0009

composition of the group. Further to the idea of the barbeque as a social framework, it is also a ritual where men have predominantly taken responsibility for cooking (Fisk et al. 1987; Thomson 1999; Skrbis 2006). In Australian culture, barbeques are firmly associated with male sociality and notions of 'mateship' and although the subject of critical reflection (Thompson 1994; Pease and Pringle 2001), it nevertheless remains an important trope in Australia's national identity, and one that is routinely reflected in WiFi group activity and discourse. As mentioned, out of 70 members in the group, only 3 were women. Although I was aware of this imbalance from early interviews, it was not driven home until I attended my first meeting and walked into a room of men. This is relevant because unlike what is well established about male dominated areas of science (Traweek 1988; Wajcman 1991; Harding 1991), engineering (McIlwee and Robinson 1992; Faulkner 2000, 2006, 2007), architecture (Fowler and Wilson 2004) and ham radio (Haring 2007), I wanted to believe the rhetoric of the network on group's website: 'Our members are people from all walks of life and ages including enthusiasts, IT professionals, radio amateurs, educators and everyday people'.

Much like burnt sausages, gender was not one of my initial lenses into WiFi technology. Yet, as this experience reveals, the 'everyday people' of whom this mission statement refers were in reality men. Unlike other volunteer not-for-profit groups such as the Australian *Men's Shed Movement* or the *Country Women's Association* which render clear the nature of their gender domains, the WiFi group emphasises 'community' and the diverse skills and interests of its membership that come from 'all walks of life'. Given the overwhelming incidence of male members this was not the case. WiFi appeared to be 'technology as masculine culture' (Wajcman 1991). Ignoring the high incidence of men in a community group that is in theory open to all genders has the effect of accepting and naturalising WiFi technology as being for men only, rather than questioning why this is the case. This is not lessened because it is 'just a hobby'. As Haring (2007) observed in her study of ham radio operators, many hobbyists used their involvement in amateur groups as training for and access to the radio profession and associated industries. Similarly, several WiFi members had gained entry into local ISPs and IT consultancy positions as a result of their demonstrated interest and developed WiFi technological skills. The number of men in the WiFi network raises a number of interesting and complicated questions: What is it about WiFi that attracts mostly men? What is inherently masculine about the

DOI: 10.1057/9781137312532.0009

way the group makes WiFi and what, if any, influence does this have on how WiFi is represented?

This is not to say women were not involved in the group. I encountered many women on the fringes of the network. Wives, mothers, sisters, aunts, female friends and girlfriends regularly helped source materials, host barbeques, pay for electricity, provide transport to events, care for children, cater for installations, source materials, agree to the use of the backyard, roof or kitchen as sites of WiFi production and otherwise sustain male makers. Although appreciated and viewed by many as part of its success, they nevertheless largely remained hidden behind the scenes. Sometimes, they would come into meetings and sit at the back with a book. When invited to join in, a typical response was 'No thanks, I'm not interested'. Yet, gradually, I learned that many of the women on the fringes of the group were interested in using the network once it was up and running in their homes.

BEN: My sister loves Air-Stream heaps. She browses a lot of people's FTPs [uploaded content] and stuff. She doesn't really get involved in the talking to other people because she's not really involved in the Air-Stream type of thing but she definitely does a lot of downloads.

Rather than locating the blame on women for being 'uninterested' in technology, many scholars have examined how technical disciplines, artefacts and representations might be shaped otherwise to accommodate women (Cockburn 1983; Harding 1991; Faulkner 2000, 2006, 2007). Faulkner, in particular, has done much to tease engineering from masculinity, asking 'how gender "sticks" to engineers' (2000:89). Like Faulkner, I did not attempt to ask 'Where are the women?' However, given the absence of women, gender became a subtext through which other actors and their actions were rendered all the more present and I began to ask 'Why does WiFi stick to men?'

Acutely aware of the imbalance, the WiFi group actively sought to lower traditional barriers to entry that inhibit women from entering IT sectors. These included holding meetings in a familiar space (local public school), not requiring costly membership or advanced technical skills to participate, providing a warm welcome, visually representing complex ideas and emphasising the sociality of the group. My involvement as a female researcher was seen by some as an opportunity to represent different aspects of the group and there was regular reference to my work on the group website. Others thought that the 'channels' through which

DOI: 10.1057/9781137312532.0009

people were 'exposed' to WiFi networking contributed to the gender imbalance. One such channel was the LAN gaming events, which have traditionally attracted young men. Given the importance of the visual culture of the group in recruiting, communicating, buying and selling goods and connecting people, it holds that attending to the way the group represents itself both reflects and produces ideas about who can and cannot participate. As Latour has written, a group's visual culture presents 'what it is to see and what there is to see' (1990:30).

Disorderly design

Ron yells from the door of the main school building that we have to start now and that we are a bit late because the barbeque has been so good. We move inside one of the single storey stone buildings. It is the computer classroom, with windows at the back of the room and a teacher's desk and projector screen at the entrance. The rest of the space is filled with four rows of desks topped with bulky beige box monitors and keyboards, black office-type chairs and a whiteboard on rollers. The white board is covered in red and orange writing from the day's class. People wheel chairs into an uneven semi-circle facing the front desk. Because there is not enough space for everyone to sit side by side, some perch on the edges of desks, between keyboards and monitors and in the aisles. The conversation in room slowly reduces and I look up from my small circle with Jan, Kerry and Kurt to see Tim talking at the whiteboard with a marker pen in his hand. I make notes about how I seem to have missed yet another start to a meeting.

Clustered conversations begun in the quadrangle continued inside the meeting room with individual threads merging into one and the meeting continuing from that point only to splinter a short time afterwards. Although an agenda was posted on the website, it was not distributed in meetings, with the exception of the AGM. Conversation was guided by Tim and Ron, yet also allowed to meander in different directions depending on who was in attendance, who was new, what questions were asked, the kinds of objects people brought with them, recent weather and its effect on signal, what work had recently occurred or other relevant news. There was a collective sense of managing the meeting shared by all present; they did it together. However, trying to gain a sense of who was who, their role and relationships was not easy when people swivelled on

DOI: 10.1057/9781137312532.0009

their chairs, literally rolled into new groups, swapped seats altogether, stood up, leant over desks, or got up to talk or look at and touch technical objects or photos on phones. It was a practice that enabled members to switch easily into different conversations, to attach themselves to stories and make connections. My field notes were punctuated with 'the conversation shifts', 'the meeting turns', 'following on', 'I overhear' and my annoyance about regularly missing the start of meetings.

Guided by ethnographic literature, I read my inability to follow these twists and turns as a symptom, along with feeling overwhelmed and regularly lost, of my ethnographic apprenticeship. It was not until almost a year later that I realised my major stumbling block was in trying to find a neat trajectory to make sense of it all. It was my narrow understanding of what should be happening that complicated my vision of what was actually taking place. It was not so much that I did not anticipate mess, as mess in ethnographic practice is expected (Law 2004; Hammersley and Atkinson 2007). I was seeking a gradually reductive process by which the messy texture of practice would somehow transform into a smooth, single and infinitely transportable series of facts. On a quest for singularity and order, I was initially blind to another way of understanding the group. As illustrated by Joel's experience, it was possible to navigate through Air-Stream's culture by 'attaching' to a myriad of visual representations. An illustrative example was the new portal.

Tim reminds us that the main point of the meeting is to talk about the new portal system that he designed in time for a recent public 'IT Discovery Day' held at a local community centre. He explains this version is customised for the event but members can configure it on a site-by-site basis. The aim is to provide a way for people to accidentally 'stumble' upon Air-Stream. It will appear on screen when a person with a WiFi enabled device enters an Air-Stream network area and will replace a stream of numbers and complicated code, which up to now has required sophisticated knowledge to decipher. Instead, the aim of the portal is to visually entice the 'stumbler' to want to learn more about the group. It will operate as a representation of, and entry point to, the Air-Stream network. It will connect people to the group. However, what Tim is talking about is not a single portal, but software to produce an infinite number of portals limited only by the imaginations of interested members in the group who each will be able to customise it to the area in which they are located. The diversity of portals offers multiple entry points into the group and the network gains in strength with each iteration. Again, far

DOI: 10.1057/9781137312532.0009

from weakening the group, differentiation and distribution is seen to make the network stronger.

Tim says he is going to 'get to the guts of it' and turns to the whiteboard, already covered in red and orange instructions. Although a cleaner sits on the lip of the board, he flips the board so the orange and red writing is inverted and picks up a blue marker. He draws, talks and answers questions from the group, which prompts new directions of diagrams and text on the board. Soon it is covered in a network of orange, red and blue lines, text and sketches.

Information does not so much flow as overlap, tangle and layer as people call things out. The group tonight is made up of 'originals', 'officials', 'tourists' and 'newbies' who are trained graphic designers, 'techies', ham radio amateurs, a researcher, computer scientists, IT specialists and other visitors with unknown interests. To contend with this variability Tim asks and answers questions and repeatedly scans the room for comments and responses. It was challenging to document the process; talk was rich and rapid and the diagram was too vivid to fully capture in flat two-dimensional mediums. Meaning was contingent on clustered chatter, visuals, noise and movement. As per its name, the demo was a live and dynamic representation of an idea.

Tim asks Simon to 'give the site a whirl' and he opens his laptop connected to the projector and a screen appears on the wall in front of us. Simple navigational boxes line the left hand side with type in the centre. Apart from the logo at the top, nothing reflects the current website. Someone says it is hard to read in the brightly lit room. Kerry switches the lights and the room plunges into a grainy darkness, lit by the glow of the screen and five laptops (Figure 4.4). Tim assures us that it is easier to read on a computer screen and the laptop users agree. He tells us we have been looking at 'the front end' and asks Simon to 'show us some of code for those of you who are techie'. Suddenly, the graphic screen is replaced with lines of code. The 'techies' ask questions and Tim stands at the screen to point out answers.

Tim's demo started with a story, shifted to a diagram on the whiteboard and then moved on to laptop computers and finally a projector screen. It featured the visual front end and then the coded back end in bright light and then in darkness. Those with laptops were encouraged to navigate the portal on their own, moving to and from pages directed and undirected by Tim. At several points, the portal was present in multiple dimensions. Dimensionality is an important component of

DOI: 10.1057/9781137312532.0009

FIGURE 4.4   *Tim's demo of the new portal on the whiteboard, projected screen and laptops*

Air-Stream representational practices. At this meeting, WiFi has taken two-dimensional (photos, maps, website, diagrams), three-dimensional (t-shirts, antennas, events and demonstrations) and non-dimensional (talk, electromagnetic signal) forms. Someone asks about how the portal will handle a lot of users and Tim suggests that people with laptops can try 'hammering it now'. The sound of typing fills the room. Tim asks Simon to go to the 'guestbook' to show how visitors can leave comments. Simon laughs and points out that 'Drift' is already there. Drift is a member who is not at the meeting but has logged into the portal via the network from his home. We see his comment on screen, 'Drift: Say hi to everyone at the meeting for me.' Everyone laughs.

Even absent members can contribute and participate in the meeting, via the portal's online guestbook, further emphasising the multi-dimensionality of the visual culture of Air-Stream. People with laptops may be physically in the room but are connected to each other via the portal, and the rest of the group divide their attention between the whiteboard diagram, the projection on screen and clustered discussions about individual designs. It illustrates Tim's method of 'visual attachment' in practice; rather than establishing a single focal point it welcomes multiple interpretations. A short time later, he expressed his thoughts on the demo:

TIM: I don't think it was ideal. Now I could do a better one.
KAT: Why wasn't it ideal?
TIM: Minor details. But as far as getting the idea across it did its job. I mean the main concept is that when you log in you gain access through the firewall and that's what I was trying to show. I didn't know that I was going to go into that

DOI: 10.1057/9781137312532.0009

much detail about all the other components. Some of the guys there would understand that, some wouldn't. I was just going to play it by ear.

The crucial point to glean from Tim's demo is how representations in the group do not glue or link people together in hierarchical sequences. This does not mean they are not sticky. People adhered to Tim's presentation on the white board, the screen and laptops. The difference lies in the fact that it is not a universal stickiness or myopic view. No one is bound or stuck to a single representation, even the designer himself. Instead, the portal works due to Tim's informal and customised 'play it by ear' approach and the group's eagerness to participate and contribute.

Spending time at the WiFi barbeque meeting has revealed some of the group's rich and diverse visual culture that shapes, and is shaped by, the idiosyncrasies of the technology, location and the nature of membership. What might be seen as a scattergun approach aids recruitment and participation irrespective of the reality of technological connection, illustrating how several ideas can survive and co-exist, creating space for both the tentative encounters of 'newbies' as well as interactions of established 'official' members. The group's representational culture is not stringently ordered. It does not impose itself on or demand conformity from those it touches in order to produce a successful technology. It is collectively made of many parts so it can be easily picked apart, unravelled and elements swapped over, new links replaced and others easily adjusted. Air-Stream's multi-faceted visual culture appears to suggest that disorder does not simply happen to the network. It is built into it. It is part of the design. This enables the group to responsively adapt to constant instability, which is the theme of the next chapter.

DOI: 10.1057/9781137312532.0009

5
Trees, Birds, Sunburn and Other Digital Interruptions

Abstract: *When you take WiFi out of 'hotspots' and into the city itself it comes into contact with a range of social, technical and environmental actors. The network is stable. It has operated for over a decade and continues to grow in size and strength. Yet, members encounter an array of interruptions in the form of trees, birds, bugs, thieves, technical complications and weather. I show how rather than ignoring or tidying them up, WiFi makers build them into the network. This chapter explores how disconnections serve to connect makers to new ideas, people and places, signalling the possibility that the group's ability to deal with constant indeterminacy and multiple realities affords it durability. Air-Stream make WiFi not in spite of interruption, but because of it.*

Jungnickel, Katrina. *DiY WiFi: Re-imagining Connectivity*. Basingstoke: Palgrave Macmillan, 2014. DOI: 10.1057/9781137312532.0010.

DOI: 10.1057/9781137312532.0010

Dealing with interruption

In the preface I described a series of thefts that resulted in the loss of core antennas from the network. In one, the owners were away on holidays and Dan, 21, was the first to discover the problem:

> I just thought it had crashed. So I just waited a couple of days to see if it would fix itself. Then I was just on my way down to the bakery. I picked up a pie and on the way home...I thought, well, it's on my way home, I'll just drive by and restart it. So I drove there, got out of the car and walked down the driveway and it was like – the routers gone!

The fact that Dan went to the site 'after a couple of days' and 'on the way home' from the bakery clearly tells us he did not anticipate theft. Instead, his initial casual response suggests that 'crashed' nodes are a mundane and commonplace event. The possibility that the problem might 'fix itself' points to acceptance of interruption. Although theft is relatively uncommon for the group, it is not the only cause of breakdown that members contend with in the making of WiFi. Instead, interruption appears to be a common theme in daily experience of the network. As argued in the previous chapter, the group has developed a vibrant multi-dimensional visual and material culture that connects people together, aids recruitment and the development of new applications of the technology. Yet the account also illustrates how the network is never fully known, controlled or experienced by any single individual. It is always partial. While one member might experience technical difficulties, another will have no issues whatsoever. How do they deal with instability and interruption? What relevance, if any, do representations of Air-Stream's network hold in contradictory conditions? What might constant interruption teach us about technology innovation and knowledge practices?

The problematic and productive presence of trees

WiFi is predicated on the visual line-of-sight between antennas in the network. This means two points need to connect to each other in order to transfer data. If anything gets in the way, the signal is interrupted or blocked. For WiFi community groups in the more architecturally dense cities of Barcelona, London, Portland and Berlin,[1] blockages between points in networks are most often caused by new buildings, which alter the physical landscape and thus reconfigure the digital landscape.

DOI: 10.1057/9781137312532.0010

In Adelaide, large-scale urban development is rare and the cityscape is relatively flat. Trees, however, constitute an entirely different matter. Adelaide's small urban centre, extensive parklands and old suburbs are characterised by mature trees. Negotiating passage over or between ubiquitous leafy blockages is an integral part of how Air-Stream makes WiFi. Unlike buildings however, the fact that trees are *never* static, but are constantly growing in size and changing in shape renders them a constant source of instability.

At the barbeque described in Chapter 4, Joel, a 'newbie', admitted he had not joined the network because of 'time, trees and apathy'. The very presence of a dense wall of trees around his house inhibited not only his signal but his enthusiasm as well. Kerry and Jan were also aware of trees in their suburb as a potential blockage to their connection. Despite trying for the year of my fieldwork, they remained unconnected.

> **KERRY:** It was winter and in our lounge we had this little cantenna, a home-made thing on a stick up near my window and it was this dodgy piece of crap and if we didn't touch it, it worked, but as soon as we touched it, it didn't work, so we left it alone. Then we get up on the roof – nothing! It was really frustrating. From our lounge there are trees and other stuff and then we get up on the roof and it's clear and we are just above trees and we couldn't get a thing. And I had at least a dozen attempts up on the roof with bigger masts and better antennas. And nah [nothing]!

Getting a 'couple of packets' meant Kerry established a connection with a nearby antenna in the Air-Stream network long enough to share short bursts of data, but it was neither durable nor stable. His account highlights the discrepancy of connection in relation to seasonal change. Even though he briefly achieved a connection when the deciduous trees shed leaves, he experienced problems in summer once the foliage returned. Thus, he achieved a 'near' connection in winter, provided he did not touch the antenna, and a 'non'-connection in summer. In addition to growing in height, trees also change in volume. The location of trees and the weather pose issues for Julie, an 'official member' and experienced ham radio operator, but her connection is further unsettled by the time of day that she tries to connect:

> It was working until the sun went down and then it dropped out and it didn't work all night and then next morning I turned it on again and it didn't work until the sun came up and in the middle of the day it also dropped out and I've being trying to figure out why that might be. I've talked to an amateur

DOI: 10.1057/9781137312532.0010

[radio] friend of mine and ... we think it's to do with something in the trees. Trees might get damp in the evenings and at night, which might cause attenuation of the signal. Whereas during the day especially at the moment they dry out and that would tend to match with the fact that during winter I couldn't really get a connection at all.

Daz, 22, a 'committee member', did not at first notice the trees in the golf course nearby, until they started to interfere with his connection to the network. He set up his node in 2004 on the roof of his parents' house, nine metres above ground level. With this height, he achieved a 'good' connection across the park to another antenna. Every year, however, the trees have grown and recently his signal has started to 'drop out'. This is not a straightforward problem. First, it 'comes and goes' so it is only occasionally problematic. Second, there is a 'good thing' about the interruption; the trees filter out 'noise' on the line from the local 'truckies' (truck drivers) who share the same radio frequency. Daz's friend on the other side of the golf course is not so lucky and regularly 'cops it' in the form of verbal abuse. This means Daz's connection is sometimes 'good', sometimes 'near' and occasionally 'non'-existent. He is also aware that this range of connectivity will expand as the trees continue to grow. At some point he will have to build a taller mast or connect to a different antenna in the network.

Reg, 26, also a 'committee member', provides another example of how trees simultaneously complicate and contribute to the network. Like Daz he lives at home and was more than aware of the dense eucalyptus bush that surrounded his mum's single storey house in a steeply landscaped northern outer suburb of Adelaide. However, rather than viewing it as an obstacle he built an antenna into one of the biggest trees. Although he had already connected to the network via an antenna on the roof, he wanted to 'go higher' in an attempt to make the connection more 'reliable'. At 13 metres in height, the biggest tree was located 50 metres from the house on an elevation 10 metres above the driveway (Figure 5.1). At the top, Reg found himself 23 metres in the air where the view across the west of the city suddenly 'opened up'. Building the antenna into the tree significantly reduces the likelihood of experiencing the same problem as Daz. The antenna will not be blocked by the surrounding bushland, because it grows *with* the tree. This solution, however, does have its drawbacks: 'It's not as rigid at the top of the tree. There is more flexibility and with the cable going up there, it creates even more loss.'

The antenna may be stabilised in the tree but the tree retains unstable characteristics; it sways in the wind. There are additional interruptions in

DOI: 10.1057/9781137312532.0010

FIGURE 5.1   *Reg's antenna located in a eucalyptus gum tree (photo used with permission)*

the form of birds, bugs and the scorching effects of the sun. Regardless of a person's status as a 'newbie' or established 'committee' or 'official' member, these accounts reveal the critical and constant role trees play in the making of WiFi. Trees cannot be coded and conveniently catalogued in the construction of knowledge and new technologies. They resist being contained, neatly cleaned up or stabilised because Air-Stream's WiFi is embedded in Australian suburbia. It also occupies spaces between antennas. As a result it encompasses a range of human and non-human actors. In Chapter 4, I highlighted how connection to the network is not a prerequisite for joining the group and nor is it guaranteed as part of membership. Now, even when it is achieved, connection is never stable. These accounts of connection – 'near', 'non', 'sometimes', 'only in the morning', 'in summer' and 'when there is no wind' – signal temporal disjunctions that members have to work around.

Sunburnt technology

Simon, Craig and I walk to the carpark to help Ron unload equipment from his car. Craig asks what he brought and looking inside the boot it

DOI: 10.1057/9781137312532.0010

was easy to see what Ron meant by 'pretty much everything'. In fact, it was difficult to see how he managed to close it. Several plastic buckets overflowed with cables, cable ties, dishes, steel brackets, modems, sticky tape, nuts and bolts, drills, hammers, saws and pliers with much more concealed beneath these objects. Ron extracted a small, discoloured modem. Pointing out the blotchy yellow markings on the once beige plastic skin he tells us that this is what a device looks like when it gets sunburnt.

Adelaide backs onto the desert and fronts the Southern Ocean. This means that the city experiences mild winters and long hot summers. With almost nine months of sun a year, antennas located on suburban rooftops, trees and sheds are almost constantly exposed to its searing strength and members are used to dealing with the force of temperatures that reach over 40 degrees Celsius. In these conditions computer plastic burns; it discolours, dries and begins to peel. Ron kept this modem on the roof for a while, it did not take long for the casing to discolour and singe. Despite how it looked, however, he said it still worked. Members shared similar stories. Craig once put a modem in a 'Tupperware[2] container' that he 'found in mum's kitchen' and then 'stuck it on the roof'. The plastic lasted for a few months but it could not cope with prolonged exposure to the sun, and eventually 'crumbled'.

These accounts highlight the environmental challenges that members deal with on a daily basis. Nothing can be done to protect computer equipment against the Australian sun over long periods of time. Instead much like the treescape, members resolve this particular consequence of the local climate by employing increasingly imaginative combinations of materials that simultaneously serve to demonstrate personal ingenuity and resourcefulness. Reg's tree antenna is a key example of this. Although he avoids the problem of the trees interfering with his connection because his antenna grows with the tree, he has to contend with a new range of challenges. Even if the antenna can be located out of direct sunlight, the heat from the sun is problematic for the computer equipment inside. To counter this, Reg used a specially purchased metal box, the type used for fire systems in buildings, which he customised beyond its original design. He drilled holes, in the side and the bottom of the box, and glued two polyvinyl chloride (PVC) plastic pipes provided by a family member who works at a winery. Because PVC pipes are very expensive he says he was 'lucky' they 'found lots of it laying around'. The top tube was chosen specifically for its shape, curving at an angle so as to provide ventilation

DOI: 10.1057/9781137312532.0010

but impede rainwater. With two steel poles, a combination of Hills Hoist (clothesline) wire bought from the local hardware shop and fencing wire that Reg 'found in the shed', he secured the box using a 'spiderweb' of chains and U-bolts into the uppermost branches. Reg's setup is uniquely shaped according to found and re-adapted materials coupled with sensitivity to the context of use.

Bugs, birds and 'gully winds'

'Bugs' are another group of actors that Air-Stream members have to work with on a daily basis. Anyone who has spent any time in Australia in summer is keenly aware of the presence of an abundance of insects. With over 220,000 different species, small things that fly, jump, crawl and bite are a mundane part of everyday suburban life. Although boxing up equipment serves to reduce the effects of sunburn and heat, members have to find ways to prevent insects from nesting inside equipment and housing. While some use sticky tape or cable ties to firmly keep box lids in place, the most popular material is flyscreen, a small gauge gauze mesh, more familiar on the swinging front door of an Australian suburban house than a sophisticated technological system. WiFi makers extend the use of this material to cover ventilation holes to keep all manner of insects out of their equipment. As Goggin has argued, it is important to 'see the novelty that different cultural contexts bring to bear on our own normative understandings and expectations of a given technology' (2007:44). Everyday domestic materials work effectively in these new applications. Equally, members are open to viewing them in new ways.

Birds are also central actors in the network. Evidence of how local fauna is accommodated in everyday life can be found in street signs installed around the city (Figure 5.2). Further to warning human inhabitants about the presence of birdlife, that 'swoop from above or below' they advise keeping away from this area at certain times and how best to avoid provoking an attack. Part of living in this city involves living with local bird-life. This goes some way to explain Tim's casual acceptance of the fact that he received a text every time a bird sat on an antenna connected to his temporary alarm system on the factory roof. Moreover, beyond bugs and birds, WiFi makers contend with other sources of climatic instability. The rugged tree covered landscape of the Adelaide

FIGURE 5.2 *Street signs in Adelaide alerting locals to the presence of wildlife in the city*

Hills has a distinct influence on the weather and wireless work. At one meeting I noticed Ron yawning. He saw me and apologised, explaining he was tired:

> The gully wind is really loud around my house. It is very scary because it is a blustery wind, and a big gust can make you tumble off a roof. Sometimes what seems like a good time to go up to the roof and fix an antenna changes when up comes the wind.

'Gully wind' is a local name for a particular summer weather condition generated from cold air rushing down the hills into the gullies and out along the plains and it is not just loud, it is also dangerous for wireless work. In Sydney, a similar wind condition is called a 'Southerly Buster' and in Perth, 'The Freemantle Doctor'. Reg's tree node was deliberately designed to accommodate local winds around his house. Using various types of wire he fastened the pole to the tree trunk and chains that looped through the box and around the branches. This method is apparently so effective he can stand on it 'like a platform'. However, despite the solidity of the setup in the tree, he has little control over the force of the wind on the tree itself. The wind makes the tree sway which in turn makes the node less reliable than if it was affixed to a building; 'Once it was

DOI: 10.1057/9781137312532.0010

extremely windy and it felt like it was going to snap. It was moving half a metre side to side.' The propensity for this gusty, unpredictable wind serves to shape when wireless work can take place and also how antennas are built and installed.

What emerges in these accounts is an ongoing relationship between technology and the specifics of the environment in which it is placed; a developing and shifting conversation between the many actors involved in making WiFi. Rather than attempting to order them into a stable, linear relationship and construct a single definite reality, WiFi makers work with instabilities, constantly adjusting and adapting to changing circumstances.

'Sometimes it just doesn't work'

Ron initially joined Air-Stream to make a connection between two of his business sites. Although he was well versed in other DiY technology cultures such as ham radio and radio astronomy, he was initially perplexed by the idiosyncrasies of WiFi.

> I thought it worked like radio. I had a good knowledge of radio. Radio goes through walls and it goes through trees and so on. If you use a UHF radio and they say the range is two kilometers and the area is relatively flat and there is no big hill in the way, then two kilometres is going to work. I assumed WiFi did the same thing. But the WiFi that we were working with could not tolerate multi-path signal. Microwave does not go through things particularly well, and it couldn't use a multi-path layer. It had to use a true original signal path – line-of-sight. You could go to someone's home, sit on the roof and go, 'Yeah, that'll work' and 'As long as we do not have too much interference that should be ok'. That's what I had assumed. I thought, well this is higher than this. In theory, that should work. It didn't work.

With Tim's help, Ron soon had it working but it was not straightforward. Mackenzie writes: 'Of all the wireless networks, most of which are not more than a decade or so old, the most unstable, the most prone to sudden proliferation or contraction, might be WiFi wireless networks' (2007:95). Because there are no wires to trace and coupled with the heterogeneous network of actors involved, there are infinite possible causes of interruption and disconnection. The concept of 'unknown' aspects of the network emerged frequently in discussion:

DOI: 10.1057/9781137312532.0010

SIMON: Everyone I've told is interested. But there is a step that goes from inter-ested to doing stuff and that's a fairly big step. Purchasing the equipment and stepping into the unknown of whether you connect or not.

KAT: Stepping into the unknown? What makes up the unknown in this case?

SIMON: You never know if you're going to be able to connect or not. The access point might be 100 metres away but there might be a big lead shield between you. Not that it's very likely but there're so many factors in it. We try and do as many tests at people's places as possible but some people just purchase the equipment and just go for it. In the case of Kerry, he's got the equipment and we went over to do a huge setup day and they just couldn't connect. All the stats said it was right, figures were fine and trees ok, but there was something blocking it. That's life. It's just that element of mystery. Because, while we know how it works we really don't know how it works. Why is that dish talking to that dish over there and I'm able to send huge amounts of data? I guess it's all electro magnetic waves, but is it really? There are some places that you say it's definitely never going to connect. There're just too many trees, it's just too far away, too many hills. And we'll get up there and point a dish and you get super signal and you go, why? There's no explanation for it but okay, we'll just accept it and use it.

Simon's version of 'stepping into the unknown' evokes the openness and mutability of technological uncertainty. When he says 'we know how it works' and 'we really don't know how it works', he reveals how he and other members are comfortable with knowing *and* not knowing the network and enfold these instabilities into their representations of WiFi.

Family involvement

Considering many antennas are located on personal houses (or trees), and given the high number of the WiFi makers still living at home, another area of stability and instability concerns family involvement. Parental or partner approval is essential to ensure use of materials, tools, the roof and backyard as well as access to and payment of electricity. What might seem a good idea to a WiFi maker may seem less so to the owners of the house and siblings. I did not encounter anyone whose family objected to their involvement in the group, but I noted many instances of social negotiation. As a result of membership being pre-dominantly male, these examples are primarily in relation to mothers, sisters, wives and girlfriends. To keep his computer equipment cool, Reg purchased two powerful fans that ran at 7000 revolutions per minute

DOI: 10.1057/9781137312532.0010

(rpm) and installed them in the tree. They proved effective for this task but they had another more unexpected feature:

> When I was putting it together, the girls [his sisters] were saying, 'Are you serious? That's bloody well loud'. It turned into a pretty funny story. You know how you get those really quiet nights? Well, my mum's room is 100 metres away from the tree and she said she could hear it.

Although the fans worked to keep the technology cool, the fact that they were located in a domestic setting meant they did not work effectively for the task at hand. Reg responded by installing a fan controller so he could adjust the speed, which addressed the problem in summer. In winter he was able to switch it off. The fact that Reg's family was unhappy about the fan noise shaped the design. Getting power to an antenna ten metres up a tree also proved to be another family issue.

> We have a power connection outside for an electric fence box. It [the power cable] runs down the tree. It's strung across two trees and across the driveway. But it's pretty damn high, so you wouldn't even notice it if you were driving down the driveway. And then it goes down the side of the paddock about 80 metres or something. Just as long as no one goes there and starts hitting the cable with a hammer or chews through it, it will be fine. But mum is worried about it... fires and stuff.

Despite the fact that Reg has solved the electricity problem, his mother's unsurprising concern about bushfires brought about adaptation in the design. Reg's next iteration involved digging a trench under the paved driveway to the electric fence in which he laid two ethernet cables and a speaker wire. This involved further negotiation with his family as it temporarily disrupted their everyday movement in and around the house. All the while his antenna leeched electricity from the domestic grid as he continued to adapt and change his design.

> I might be doing something and her friends will come over and say, 'What the hell is your brother doing up a tree?' 'Is he like Spiderman or something?' And I laugh.

Reg's family's support in terms of resources and good humour, considering the interruptions his activities cause, illustrates how the network is deeply embedded in existing domestic ecologies. Reg's mother's support is essential to the success of his antenna in the network and his sister's jibes serve to fortify his innovative and experimental identity. Although women, and more broadly families, are pivotal to how the network is

DOI: 10.1057/9781137312532.0010

made and maintained, they are not as visible as other actors in representations of WiFi.

Feminist STS scholars have long drawn attention to the lesser-known roles of women behind the scenes of technological systems (Schwartz-Cowan 1983; Star 1991; Wajcman 1991, 2004; Cockburn and Ormrod 1994). As Wajcman writes, '[T]heir absence is as telling as the presence of some other actors, and even a condition of that presence' (2004:41). This scholarship draws attention to women's voices, experiences and perspectives and in the process raises questions about men's monopoly over the history of technology. In the past, few women's voices have been heard because their contributions have either been rendered invisible or not recognised as important, with many technological inventions appearing to come from a single (male) author, reinforcing the romantic ideal of an independent inventor. In reality, it was more likely groups of people who worked on inventions, both men and some women. Moreover, as Schwartz-Cowan points out, it is not just some inventors we hear little about. It is also certain kinds of technologies:

> The indices to the standard histories of technology ... do not contain a single reference, for example, to such a significant cultural artifact as the baby bottle. Here is a simple implement ... which has transformed a fundamental human experience for vast numbers of infants and mothers, and has been one of the more controversial exports of Western technology to underdeveloped countries – yet it finds no place in our histories of technology. (1983:52)

Reflecting on Schwartz-Cowan's argument, an even wider range of makers would tell even more diverse stories about the wireless network, further expanding what it might be used for and whom else it might attract.

Volunteers, time and skills

'It'll make a big impact on the city. It will open up the south and double the network.' Peter's comments elicit excited murmurs in the group at a monthly meeting. Ron interrupts, curbing the enthusiasm a little. 'Yes, but it only gets done if people do it. It's only as good as the members who make it and sometimes people get busy. It's up to all of us to get it up and running.' Without members' skills, bodies, houses, money, time and financial investment as well as the often overlooked, yet essential support of their

DOI: 10.1057/9781137312532.0010

families, the network would not exist. As illustrated at the barbeque in Chapter 4, the group is primarily made up of men aged 18–25 and 35–50. This age gap is aligned to life stages. When members get to a 'certain age' they get into a relationship and 'disappear'. Similarly, the 'middle range' members get 'busy' with jobs or family life. They 'discover' or 'return' to the group when they have time again. While a shifting membership of this nature suggests potential destabilisation, in the context of Air-Stream it means a diverse group shares the financial, social and technical burden of the network. Tim and Ron provide good examples:

TIM: We don't sleep. We think of doing something at 10 o'clock so we just do it.
RON: Those with families and things have to organise stuff during the day.
    They [young people] haven't got children to put to bed.

Younger members are likely to do emergency, evening or weekend work and invest more of their disposable income while those with family commitments have connections to businesses in good locations, materials and tools and frequently provide car rides for younger members. Older members need more time to plan their involvement and have periods when they cannot do wireless work. Craig and Peter provide another example. At 20, Craig is one of the youngest in the group and, at the time, one of its significant financial supporters. Single, living at home and working in his first job means he has more flexible time and fewer financial commitments than older members with families. He was quite happy to be on-call if something broke because his 'dream for years' has been to 'get everyone connected'. Peter, 40, employed as a social worker and married with children, viewed his participation to the group in a different way. During an installation he explained how he did not have a disposable income like Craig who is 'really passionate' and committed to it 'every waking moment', but he was more 'interested in people'. He also had a work car with access to subsidised fuel. 'So that is my contribution. I carry around gear.'

Given the diversity of membership, contrasting motivations could signal instability. Yet, the group works with and around them. Large complex installation events that require lots of volunteers are not organised around long weekends or public holidays. Likewise, some of the younger members are not available during university exam times or when music festivals come to Adelaide. The heterogeneous and disordered nature of the group means that someone is always available to do wireless work. In this way, an unstable membership contributes to the stability of the group.

DOI: 10.1057/9781137312532.0010

Moving house, moving antennas

The short-term nature of rental property poses an ongoing challenge for the group. Kerry and Jan had trouble installing an antenna on the roof of their rented property. They were penalised for damage allegedly caused by Kerry in his attempts to get connected. As with Jason's story at the barbeque, antennas that go 'down' do not always find their way back 'up' again when members move. House or office moves, however, do not always threaten destabilisation to the network. Not everyone dismantles their antenna when they move which results in the installation of another antenna at a new site. Sometimes, a new site is more advantageous to the network than the old one. A member on the website exclaimed: 'Let's hope the new office is some sweet plum location!' Moreover, not all antennas are located on private houses. Some sites, such as a decommissioned water tower, the rooftop of a government health building, office buildings, shopping centres, university buildings and local community centres come about as a result of members' personal and professional contacts. Although these sites provide network expansion possibilities, they come with their own instabilities. The account of theft for instance provides an illustrative example. While the antenna was left intact when the WiFi member sold the house, the theft compromised the site on multiple levels. First, the group lost prime equipment and second, the new owners became worried. Rather than risk another attack on their property, they asked for the remaining equipment to be removed. The group lost equipment and a site.

This raises the issue of broader public social negotiation. Members need to obtain permission and access to the site, which necessitates advance preparation and crafted presentation. Tim explains what happened at one particular site:

> It turns out that we got permission from the wrong person. The person that we got permission to put this up on the building wasn't actually the owner of the building at all. It was some old lady down the road. She was saying, 'No worries, you're a lovely young boy, you can do whatever you want on my roof'. But she didn't realise we were referring to a completely different building. Yeah, we got a nasty letter a week later. 'What are you guys doing with all this stuff on our roof?'

Although antennas that move can temporarily destabilise parts of the network, new members and new locations play an important role in

DOI: 10.1057/9781137312532.0010

the expansion of the network. Reg continued to plan further upgrades and changes including the use of solar power in an effort to secure an even more 'reliable' connection to an antenna in the west of the city. The use of the word 'reliable' initially appeared paradoxical in these contexts considering the inherent network instability. Yet, it reveals how members constantly negotiate the desire for stability while remaining open to experimentation. During fieldwork, I heard many comments that reinforced the idea that WiFi is never finished, such as when I asked Simon what happens when they all get connected: 'We'll just keep upgrading, faster and faster'. What drives many WiFi makers is the idea that the potential of the network is never fully realised. It remains suspended in a state of making. It can always be something else, something better, faster, stronger and more reliable. Circuitously, the relentless drive to improve the network also produces constant instability in the network.

Building interruption and 'love' into the system

A poster at an Open Day reads 'Air-Stream: The *Other* Broadband'. Air-Stream is a networked infrastructure not dissimilar to commercial broadband internet in that it operates by connecting individual antennas to create networks that enable people to transmit *and* receive data signals. As a result, there is no one central hub of the system. A series of 'backbone' antennas transmit data at full speed to and from key points and distribute it between their attached clients. Each can support up to 30 simultaneous connections to smaller 'client' antennas and Dan's was one of these, which explains why the theft interrupted only a few members. The network is deliberately constructed to accommodate variability of connection.

> RON: We get the next level of member who wants to set up access points and have them, what we call, backboned to other access point and that's the redundancy system we were talking about. So if you have an access point and people connect to that then if you can provide two points and preferably three points of connection of other networks then if that network fails it can go through that.

Redundancy is built into the system. Volunteer community wireless groups around the world differ according to technical legalities, landscape nuances and their vision of connectivity shapes their infrastructure and in turn their visual culture. Many groups, such as Île Sans

Fil (Wireless Island)[3] in Montreal install 'hotspots', in a similar style to commercial providers, in local cafes, parks and community sites. They are less concerned with networked infrastructure and more with a series of distributed points through which people can access the internet for free. Fundamentally, Île Sans Fil and ISPs both deliver the internet via a central pipe to an individual user. People connect to the internet first and then to others. However, if the exchange through which the internet is piped crashes, then the connection is lost, and people become disconnected to one another until it is fixed. The hub of the system is central to connectivity.

Air-Stream's *'other* broadband' is different. It uses a distributed network that supports interdependent relationships between many clients. There is no one central hub in this system, or any limit to how far it can expand. Instead each antenna is 'backboned' into the network via multiple dedicated paths or pipes. Using these dedicated channels backbone antennas transmit data full speed to and from attached clients. This pattern is repeated with the next closest antenna and so forth, which means a person in the south east of the city can connect to someone else in the north-west. With this infrastructure Air-Stream connects people to one another, not just to the internet. Rather than copying the architecture of the internet, they are creating alternative ways of joining people together. They do this by deeply embedding antennas in the local landscape – on member's houses, backyard sheds, factories and office buildings. This is how Air-Stream members make a local version of the internet.

What this means is that the reality of breakdown, mistakes and malfunctions are built into the Air-Stream system. Yet, they are not considered interruptive in the traditional sense of the word. Interruptions do not *happen to* the Air-Stream system, they are *part of* the system. There is space to experiment and try new things, to fail, which provides opportunities for members to learn about the technology and customise the system. The study of failure is important in STS (Star 1991; Latour 1996). In his analysis of Aramis, a personal rapid transport system in Paris, Latour (1996) links its failure to the fact that no one cared enough about it to guarantee its survival. For Latour, a technology needs to connect to existing human and non-human networks, which enable it to become stable and therefore durable. In other words, a technology needs to be constantly loved.

To operate a WiFi network requires a similar constellation of emotional, financial and material investments. The success and durability of Air-Stream lies in the constant care and sustained commitment of a small group of vigilant users combined with the distributed nature of its technological design. Because the network is never considered completed or finished, it needs relentless monitoring. The distributed network of antennas means that if one breaks down or is stolen, only smaller ones like Dan's are disconnected. For everyone else, data is re-routed away from this point. This enables Air-Stream's network to continue to work in the presence of absent or dysfunctional points and means that a crashed node in the north of the network has little, if no, impact on the south of the network.

These accounts give the impression of inherent instability, which could be read as threatening the durability and longevity of the network. Yet, members do not attempt to suppress or erase unstable actors but rather work with them, making them as much a part of their toolbox as the modems, computers and wires that technically comprise the system. Drawing attention to constant challenges highlights their ability to resourcefully respond to potential problems. Anything that goes wrong can be fixed, even if temporarily 'unknown', thus defusing the threat to destabilise the network. Disconnections make new connections – with people, ideas, the city and a range of human and non-human actors. In this context, interruptions become linked to innovation and experimentation, a feature echoed in the structure of the network itself. Reg's tree antenna solution is a key example. Placing an antenna in a tree is not a problem-free solution. In many ways it manifests new and unexplored territory. The tree sways in the wind, there is increased possibility of fire from the electricity cables and rain and the box is more susceptible to bugs and birds. Yet, these unstable actors are folded into the 'unknown', 'mystery' and challenge of making WiFi, which provide opportunities for new forms of expression and technological imagining. Along the way, WiFi makers display their ingenuity in overcoming diverse impediments. It is therefore possible to consider that the group's innovative spirit might not be as vigorous or demanding without the struggles to connect and the constant threat of disconnection.

Notes

1   See http://consume.net, http://bristolwireless.net, http://personaltelco.net, http://cuwireless.net, http://friefunk.net

DOI: 10.1057/9781137312532.0010

2   Tupperware is the name for a range of popular kitchen storage products made of hardwearing durable plastic.
3   I have chosen this group but I could have selected any number of community wireless organisations that operate in a similar way – such as NYC Wireless in New York (http://www.nycwireless.org), CUwin in Champagne Urbana (http://www.cuwin.net) or Portland Telco in Oregon (http://www.portlandtelco.net).

DOI: 10.1057/9781137312532.0010

6

Representing Digital Noise

Abstract: *This chapter focuses on 'stumbling', a routine technique employed by makers to look for and represent digital noise. I describe how it constructs a version of suburbia without fences, houses, roads or power lines and argue that stripping away familiar and mundane symbols of power and ownership serves to collapse distance between people and infrastructure, reconstructing in its place an uncertain digital landscape that relies as much on social cohesion and technological imagination as on hands-on technical skill. This landscape however is not neutral or empty. Upon erasing some actors, others become visible. Analysis suggests stumbling attempts to represent a feral version of WiFi and that this (local) lens reveals power dimensions within these shifting invisible landscapes.*

Jungnickel, Katrina. *DiY WiFi: Re-imagining Connectivity*. Basingstoke: Palgrave Macmillan, 2014. DOI: 10.1057/9781137312532.0011.

DOI: 10.1057/9781137312532.0011

Stumbling in suburbia

Ben takes three steps and stops. He raises a thin pencil shaped antenna above his head. He takes another four steps and stops. Turning the antenna in one hand in elegant figure eight swoops, he pauses and swoops again. The laptop remains steady in the other. He reaches the fence, turns, takes more steps and pauses. He continues to traverse my front yard in a zigzag fashion, waving a series of antennas, one after another. Rarely does he look away from his laptop screen. Ben is not looking at my yard but rather what is in my yard made visible through his computer.

Like many front and backyards in suburban Adelaide in the height of summer, mine bore the visual signs of grade three water restrictions. Due to chronic water shortages, sprinklers were banned and only hand-held hoses could be used twice a week for three-hour periods before nine in the morning and after six at night. There was no water to waste and by mid-season, the drought-ridden grass bore a striking resemblance to my dusty cement driveway. Despite the relentless heat I, like many of my neighbours, still used the yard as an extra room, extending my domestic footprint all the way to the fence. The recycling bins by the driveway, freshly washed clothes hanging in the walnut tree and the presence of a fold-up chair point to some of the ways it was regularly inhabited and used. The actions of Ben, who came to visit one Sunday afternoon, however, did not fit with any of these activities (Figure 6.1). They reflected something else. Ben's movements mimicked those of a high-tech water dowser. He *was* searching for something in the surrounds of my house.

FIGURE 6.1   *Stumbling for 'wireless noise' in the yard of a suburban house*

DOI: 10.1057/9781137312532.0011

But it was not water. What he was hoping to find was just as precious and hidden to those without knowledge and appropriate equipment. He was 'stumbling' for 'digital noise'.

Before I explain how it is possible that Ben might 'look' for and 'see' wireless 'noise', I need to step back to explain why he was doing it. I expressed interest in getting connected to the network at a monthly meeting. Several members asked me where I lived and in what type of dwelling. When I told them of my location on the northwest fringe of the city's public parklands and the structure of my single storey brick and stone house typical of the area, many shook their heads. Even though central Adelaide is primarily flat, I learnt my suburb lay in a ditch and my house being low to the ground meant the chance of 'seeing' a nearby node in the area was slim. As outlined in Chapter 5, buildings, trees and other topographical nuances can inhibit the ability of individual points to 'see' each other in order to share data. Given my situation, I was surprised by the comments that followed:

> 'But, you might be able to see something.'
> 'You could be lucky.'
> 'You should do a stumble.'

'Stumbling' is a regular technique used by members to detect and render visible the invisible wireless spectrum or 'noise' that inhabits the air around a particular site. It is an activity employed at new sites, to fine-tune connection at existing ones and to upgrade equipment. Specifically, it involves measuring the microwave signals that emit from wireless devices. From the resulting textual and graphical data, members can determine the location, direction, strength and quality of local signals and armed with this knowledge, ascertain what kinds of wireless work would be required to get new antennas connected to the network. This includes what type of equipment is needed, at what height to install it and the optimum direction it should point to connect to the network. Importantly, you do not have to know if you can connect before stumbling, you can simply 'have a look around' which is what Ben offered to do. Specifically, he was looking for other Air-Stream antennas, which according to the group's Network Node Map were located nearby. 'There's always a chance', he said, echoing the sentiments of members at the meeting. The use of the words, 'might', 'could', 'should', 'lucky' and 'chance' suggest there is no such thing as *no way* of getting connected. There is always a *way* and in most cases, several ways.

DOI: 10.1057/9781137312532.0011

Stumbling on purpose

Ben and I had arranged his visit by email. He said he might 'drop by' on Sunday afternoon and the event was constructed as nothing special. He arrived at my house by car after 'a big weekend' with friends near the beach. He was tired but it was 'okay' because he had stumbled 'heaps of times'. He told me he always carried his 'stumbling kit' in his van 'just in case', thereby assuring me there had been no special preparation for the visit. In fact, he almost made it sound boring and talked of the visit more as an opportunity to see some of my bicycles (see Jungnickel 2013). The fact that Ben, and others in the group, saw stumbling as a mundane practice transformed it into a compelling object of study. As Garforth notes, '[T]he boring work, the routines, the manipulation of machines, materials, and texts is often precisely what the STS researcher wants to see' (2011:272).

Together we unloaded Ben's van on the dry grass in my front yard. His 'stumbling kit' or 'rig' included a black plastic toolbox with several compartments filled with a computer mouse, soldering iron, cables, wire cutters, spanner, cable ties, pens, paper, motherboard, boxed modem, three rolls of electrical tape, a box cutter and various loose nuts and screws. His fully charged laptop was installed with *NetStumbler*, an open source software designed to detect and represent wireless noise in textual and graphic data. It would reveal the name of a surrounding signal, which channel it was operating on, speed, vendor or maker of the device if applicable, type of antenna, whether it was locked or open to public use and the signal strength. Ben also brought three antennas. The largest was a black dish by a local manufacturer, with a split down the centre and held together with black electrical tape. Ben was careful to show me how to hold it at a certain angle. It was no longer part of a 'proper setup' because it was 'kind of broken' but it still 'did the job'. The particular job this antenna did was 'stumble vertically'. The second antenna, a small black stick about twenty centimetres long, was a mobile car antenna with a circular magnetic base designed to affix to the roof. Ben, like other WiFi group members, used it to 'wardrive' around the city looking for wireless signals. In the context of my yard, this small stick antenna was to be used to 'stumble horizontally'. The third, and by far the most unusual looking antenna, was a 'cantenna'. Ben made it using half a 'milo tin' [a chocolate powdered drink]. Called an 'omni', short for omni-directional antenna, it was to be used to scan the air in

DOI: 10.1057/9781137312532.0011

multiple directions. Covered in a thick layer of rust that came off on our hands, and peeling yellow electrical tape that bound the jagged tin edge to prevent it slicing fingers, Ben confirmed 'it works better than it looks'. With the cables untangled, he prepared to connect the first antenna to the laptop. This unlikely looking ensemble was thus transformed into an operating 'stumbling rig', which would be used to detect and render visible and audible the invisible wireless spectrum or 'noise' that inhabits the air around my yard.

Stumbling is defined as a mistake, a trip or to walk unsteadily. Not as dramatic as a fall, it is often considered accidental, of little consequence and easily forgotten. Unsurprisingly, it is known by more than one name in the group. Stumbling is interchangeably 'sweeping the sky', 'scanning' and 'site surveying'. All are machinations of the ordinary and the extraordinary. Just like the notion of stumbling on purpose, sweeping is an otherwise unremarkable everyday domestic practice of cleaning that becomes extraordinary when performed on the sky. Further to the absence of a single term or distinctive definition for stumbling, there is also no definitive stumbling rig. Stumbling can rely on a single device, sometimes an assembly of devices; it can be as small as a hand held computer or as large as a number of differently configured antennas connected to a laptop. In many cases, such as Ben's, a rig may comprise of an assemblage of borrowed, broken, hand-made and re-purposed devices.

Although stumbling is a mundane and everyday activity to Ben, it is not a haphazard series of events. It involves an orchestrated assembly of time, place, constellation of objects and specialised skills. Yet, it concurrently retains elements of ambiguity as Ben explains, it is also about 'getting out there and trying something which you don't necessarily know if it's going to work'.

Reconfiguring suburbia

Ben calls out what he can see as he zigzags across my front yard:

> 'Microwaves.'
> 'Cordless phones.'
> 'Linksys.'
> 'Madhouse.'
> 'Ovingham.'

DOI: 10.1057/9781137312532.0011

Using his voice, body and various devices, Ben makes known the digital contents of my suburban yard. While *NetStumbler* renders this data in graphic and text form, he makes it physical and audible. Electromagnetic waves from my neighbour's oven and cordless phone seep into my yard. Similarly present are overlapping domestic wireless broadband networks. Two are my neighbours' and one is my housemate's. According to Ben, my yard is 'digitally messy', but not as messy as other places he has stumbled. He tells me he regularly finds 'noise' transmitting from baby monitors, garage doors, cordless phones, ham radio operators, walkie-talkies and even heart monitors. In fact, there is so much wireless noise in the city that according to the group's website if you do not detect anything then 'chances are your stumbling rig is not working properly'. In line with this, the task shifts from seeing wireless noise to deciphering it. More specifically, stumblers must be able to disentangle a usable 'signal' from 'background noise'.

Being able to distinguish a signal from noise relies not only on special skills and devices but on a sense of ease with the digital presence that blankets suburbia. British designer Anthony Dunne's (1999) *Faraday Chair: Negative Radio* brings the opposite to light by rendering visible the rarity of 'empty space' devoid of electromagnetic spectrum. His installation comprises a glass box coated in conductive ink in which the user lies curled in a foetal position and breathes through a tube. The power of Dunne's work stems from the difficulty of escaping from the pervasive and threatening presence of something that cannot be seen. The novelty of stumbling lies in the visualisation of this invisible infrastructural web that inhabits suburbia and the way members attempt to domesticate a largely uncontrollable wireless landscape into everyday life, just as fences, electricity cables and driveways form the infrastructural patterns of every life. The fact that they render it visible for the purpose of establishing new antennas and ultimately expanding the network provocatively presents it not as a threatening force from which individuals need protection, but rather as inviting new ways of thinking about infrastructure and the promise of connectivity.

As Ben walks around, he narrates radio signals that seep through walls and slide through fences into my yard. Although wireless devices can be locked to prevent unofficial access, they cannot be held in one place. Radio signals slip in and out of houses and through walls and fences to occupy, mostly uninvited, new territories such as my front yard, where they lose strength in the process and tangle with others. Ben tells me how

DOI: 10.1057/9781137312532.0011

they overlap in the corners of my yard. Operating on a shared spectrum means signals can interfere and even cancel each other out. In other work I have written about how engaging with wireless technologies in domestic contexts 'predicates an understanding of what they attract and repel, where and how these overlaps interact, what is displaced and what is revealed' (Jungnickel and Bell 2008:275). With WiFi, the logics of space and place are reconfigured from a series of fixed and ordered entities of more traditional infrastructural systems to a constellation of messy contingent ones. While my previous work was located in the home, what Ben is documenting is a form of digital leakage outside the home.

Stumbling does not present the common vision of mundane and much maligned suburbia (Elder 2007). Instead it provides a heightened awareness of, and access to a digital version. Seeing *into* my yard in a way not possible before presents similarities to the advent of X-ray, which provided unprecedented access to the interior of the body, and was generative of 'new configurations of the body' (Cartwright 1995:107). With Ben's help I viewed the names people had given their wireless systems and the types of devices they were operating, which when pieced together provide insights into the technologies that give shape to their daily activities. Cartwright points out the 'perverse spectatorial pleasure of X-ray researchers and the public confronted with the static X-ray photograph' (1995:108). While previously this kind of pervasive scrutiny of suburbia was only possible by large corporations with centralised panoptical systems, Ben's actions demonstrate the relative ease by which WiFi members can bypass conventional data entry and exit points.

Despite finding a cacophony of digital noise, Ben shook his head and muttered unenthusiastically. He cannot find what we are looking for in the digital composition of my yard. However, just because we have been unsuccessful so far does not mean that I cannot get connected to the group's network. Ben says we need to 'go higher'.

Landscapes of possibility

'Going higher' offers the possibility of stumbling into more wireless noise that would otherwise be interrupted by the architecture of the house, trees and other large buildings (Figure 6.2). Using a wooden ladder from the garage, Ben shimmies between the sunshade and the overhanging veranda at the front of the house and onto the corner of the

FIGURE 6.2   *We 'go higher' (onto the roof) using a 'rig' comprised of broken, re-purposed and homemade devices*

roof, carefully navigating the connecting gutters and flashing. I pass him the laptop and two antennas, which he briefly balances on the lintel over the garage before scampering up the roof. I follow with the remaining antenna and my camera, taking slower and more cautious steps along the dry and dusty ridgeline that is hot underfoot. I note that the tiles are not terracotta, but pressed aluminium which renders them even more fragile. A tell tale trail of dents across the roof is evidence that we are not the first ones up here. We can see in detail the journey of a heavy-footed digital TV installer who bolted a bracket and dish to the brick chimney. Ben follows my gaze and points to the dish, saying 'it's a shame' the dish is not installed higher up because 'we could've used that'.

Ben's comments suggest the roof offers yet another digital landscape of possibility. In addition to looking for wireless noise in spaces between antennas, WiFi makers are also often on the lookout for materials that can be incorporated into the network. Existing roof infrastructure, in the form of television antennas and even chimneys, are regularly built into WiFi setups. An example is how Reg incorporated an antenna into a tree located on his parents' property that had been blocking the signal from his house. It was a tactic that significantly reduced the chance his signal would be hindered by local flora in the future because the antenna literally grew in height with the tree. Members also regularly talked about scavenging materials from 'hard rubbish' (monthly collections of neighbourhood white and brown goods on the footpaths prior to Council pickup) and I regularly noted the use of re-purposed domestic objects

DOI: 10.1057/9781137312532.0011

such as clothes line wire, plastic Tupperware containers, biscuit tins and fly screen. The suburban yard, house and rooftop not only provide key sites in the network, they are also integral in the making of WiFi.

Ben looks around using his eyes before he looks with the antennas. Up here yet another version of suburbia is evident; a mottled landscape of fenced quarter acre blocks, rust streaked corrugated iron roofs, trees, backyards filled with fruit trees, chicken runs, vegetable patches, outdoor sun shades, sheds, children's toys and (mostly empty) in-ground pools. The soundscape on this Sunday afternoon included yapping dogs, lawn mowers and squealing kids. Height is important because WiFi is predicated on the visual line-of-sight between antennas in the network. As discussed in Chapter 5, two points need to connect to each other in order to transfer data. If anything gets in the way, the signal is interrupted or blocked. WiFi makers talk about the importance of being able to 'see' from point to point. While this refers to the visual and the technical, it also involves the imagination.

Seeing the unseeable

Ben stumbles on the crown of the roof in much the same way he did in the yard (Figure 6.3). In one hand he holds an open laptop, with the other he loops the antennas in lazy horizontally and vertical circles. He squats to change the position of his feet to stumble in the other direction. He keeps his eye on the laptop, looking for wireless activity. Every few minutes he bumps the keyboard with his nose to interrupt the screen saver. His movements are slow and sure. The only difference to the yard is the fact we are perilously perched on the crest of the roof located over five metres above the ground. We may only be on top of a single storey house, yet I am keenly aware of the height. The roof tiles slope to the edges of the house and there is nothing to stop us from falling if we were to accidentally start to slide. The fact that it is a really hot afternoon does not help. I'm sweating and my skin picks up dirt and dust from the tiles making my hands and feet red. Instead of making it easier to grip, my dusty bare feet slip on the tiles and my fingers are splayed in a telltale grip of fear on either side of me. I am torn by my split desire to hold on to the roof and let go in order to take photos. Ben does not seem to think this is a particularly steep roof. He has 'been up much worse' and entertains me with stories about some of the 'worst stumbles' involving

DOI: 10.1057/9781137312532.0011

FIGURE 6.3   *Ben stumbles on the rooftop of a single storey suburban house*

lightening, rain and wind storms, slippery tiles and extreme heights but I cannot remember them in detail because I am caught up in imagined chaos of broken limbs, trips to casualty and how I might explain to the owner of the house if one of us falls through the roof.

Reflecting on this later, I was struck by how comfortable Ben was on the roof while I was less so. This was at odds with my other experiences such as a recent decommissioned water tower installation where, as an WiFi 'dogsbody', I climbed thin steel rung ladders between double height floors and helped wind overflowing milk-crates of equipment through an unfenced cavernous open central core. Yet, this experience revealed something else. Ben treated the roof just as he did the front yard. He wandered around, looping the laptop and antennas around his head, all the while talking of what he could and could not see. The roof was so ordinary to Ben that he barely noticed it. Dunbar-Hester (2008) found herself in a similar situation during her ethnography of American 'geeks' primarily engaged with low-powered FM radio but who also tinkered with WiFi networks. She observed and participated in a range of hands-on activities including holding the legs of a WiFi maker as he hung upside down from the side of a house.

> Events that contained an element of danger were not performed with an overly dramatic flair, but at the same time, I argue that the display and management of risk in some of these settings (working with and stories about high voltage/current, power tools, and heights) did include some masculine bravado. (2008:213)

Ben's actions could similarly be read, however I suggest two other reasons as to why the dangers of the rooftop disappeared. The first relates to Cartwright's (1995) account of Edison, Dally and other scientists who sacrificed their own bodies in gruesome experiments with early X-rays. She argues that what drove them to 'pursue a technology that demonstrated so clearly its potential for bodily destruction and death was not only the thrill of seeing the deathly spectacle of the skeletal system but also the potential to harness the physiological force of the ray as a medical treatment' (110). Although, not as grisly as the effects of experimental X-ray, nevertheless the danger of the rooftop is similarly offset by the thrill of seeing the unseeable. Danger is trumped by what is revealed.

Another interpretation involves a cultural and embodied reading of the Australian rooftop. Having lived in central London for ten years prior to this study, I realised I had forgotten about rooftops as domesticated spaces. Briefly analysing this space, and my response to it, reveals the cultural edges of the everyday and points to the distinct character of the Australian wireless landscape. As noted by Hall (2007), Adelaide is predominantly made up of single storey houses on large residential blocks. This means that rooftops are relatively accessible from the ground, are more likely to be individually owned (or rented) and as a result considered extensions of domestic space. Like the yard, garage and kitchen, rooftops are sites of daily chores. For instance, roofs have long held television and radio antennas. When storms strike these infrastructures require someone climb up and adjust anchors that hold it in place. Gutters also need attention on a regular basis. With the worsening drought, raintanks are no longer the preserve of remote homesteads and are increasingly found in suburbia. This means that gutters do more work than simply channel water into drains. They provide a life support system and need to be regularly cleaned of leaves to ensure they work efficiently. In times of bushfire, gutters are enrolled in fighting fires. Emergency services advise residents to wedge tennis balls in drainage pipes and fill them with water to prevent flying sparks erupting into spot fires. Rooftops are also prime surfaces for the collection of energy with the installation of solar panels and many houses also make use of ceiling

DOI: 10.1057/9781137312532.0011

skylights for natural lighting. All of these activities require regular maintenance and support the idea that the roof is a site of domestic tasks and responsibilities. The fact that a neighbour I did not know, who spotted Ben and I on the roof with what must have appeared to be a curious collection of devices, cheerfully waved at us instead of calling the police also points to a cultural familiarity with rooftops.

WiFi and the technological imagination

In the kitchen, over beer and peanuts, Ben shows me what he saw in the yard and later on the roof. His laptop sits open on the bench revealing text and graphical data. My laptop is open to a map of the area and a printed Network Node Map lies between them. Ben explains the results via the names, direction and strength of nearby signals. Although there was a lot of 'noise' outside the house, Ben tells me there is nothing I can use 'at this stage' to connect into the network, but assures me this does not mean I cannot get connected. There is still 'a chance' and suggests I install an 'omni' as high as I can to 'see who can see me'. Throughout the afternoon Ben stressed how stumbling from my house produced a unique view of the network to other sites and that if he was to stumble here again in a month, a day or even in an hour, it would result in a whole new series of data. Even if we saw a strong signal from a nearby antenna, getting connected was still not guaranteed; the view of the wireless landscape can only ever be temporal and partial. He reminded me that there is nothing immutable about stumbling. Although this practice produces data about digital noise around my house it does not fully capture nor fix what is out there. It cannot hold it. Paradoxically, Ben wanted me to believe in the process of stumbling and the information embedded in the representations and to simultaneously disregard it. He did not want me to give up hope that I could connect to the network, even though the stumble clearly indicated otherwise. By suggesting I install an omni-directional antenna despite evidence to the contrary, Ben was encouraging me to embrace the ambiguity of the technology and to imagine myself in the network. So, how do we make sense of such contradictory representations of knowledge? What use are they in the context of building a new technology?

Cartwright describes how the X-ray was seen as a 'wild, unknown natural force that had to be harnessed and managed in order to be put to

DOI: 10.1057/9781137312532.0011

good use' (1995:110), was institutionalised 'as a form of diagnosis' and 'as a powerful means of disciplining the body' (1995:123). It was represented in terms of order and control. Scientists imposed themselves upon it to make it work successfully and then imposed it upon the body, often their own (or in many cases, women's bodies). Stumbling produces representations that are similarly uncontrollable, however Ben does not subjugate or attempt to harness wireless noise, nor does he interpret the results in terms of success or failure. He does not impose himself upon the wireless spectrum but lets it reveal what currently exists, imagines what might be possible and sets about weaving the potential of a new antenna into the social fabric of the digital landscape. The practice of stumbling in this case is less about mapping a landscape or diagnosing a problem and more about opening up a range of possibilities.

Like the X-ray, stumbling penetrates the surface of the suburban body and opens it up to view in ways previously not possible; it produces snap-shots of a particular moment in time. Similarly, it can only be understood in relation to social and cultural contexts – the person who makes and interprets the data. However, unlike the X-ray, stumbling is not a one-way, linear process: it is dialogic. When Ben suggests I should 'see who can see me', he reminds me that I am not the only one with access to the digital landscape. Equipped with appropriate skills and tools, the network can be viewed from multiple directions and at different times. Interpreting these representations of the digital landscape, Ben constructs a vision of the network that fits with my predicament. He is not led by the data, but rather shapes it into a series of responses. As Garforth has argued, '[I]n/visibility is not simply a matter of what is (not) there to be seen' but rather seeing is a process of subtle social and cultural negotiation that is 'dynamic and practiced' (2011:266). I noted other examples where WiFi was inscribed with chance and possibility.

REG: I set it up not knowing if I'd be able to connect but just out of interest but with the hope that one day I'd be able to connect somewhere.

RON: Just because you can't see the network doesn't mean that you can't connect to the network. It's about finding other people who can and getting them [connected] and then connecting off them.

WiFi makers do not employ stumbling simply as a way of representing a single reality, but as a means of expanding the network through a technological imaginary. It provides information that enables the user to 'see' the digital suburban landscape not for the purpose of producing

DOI: 10.1057/9781137312532.0011

direct answers but of crafting possibilities and developing strategies for social and technical connection. The question is not 'Why can't I get connected?' but 'How can I get connected?' Far from closing or narrowing alternatives and choice, stumbling serves to open up multiple and dynamic expansive social landscapes of connectivity. It reveals a dynamic landscape in which not knowing is rendered productive and sticks people together, regardless of the status of connection. Stumbling illustrates how being un-connected is a valid position within the group because there is always a hope or chance of getting connected.

Building collaborative connectivity

I observed other examples of the group's collective technological imagination at work. Although a 'newbie' to the group, Kurt was highly experienced in IT. Introduced at the barbeque meeting in Chapter 4, Kurt was aware that he 'lives off the map', meaning there was little chance he could see the network from anywhere near his parent's house where he lived. Yet this had not stopped him imagining other ways to get connected:

KURT: I used to be able to see Mount Barker from the top of the shed. Until the trees grew I would have been able to bounce [a signal] off it, if we could get wireless gear up there. That's assuming I only run gear off the shed because the house already has [Digital TV] gear on it and if I was to take that dish down and potentially realign it and put it somewhere else on the roof, I'd be able to bounce off the house over to Mount Barker. Or I could get a higher line of sight to someone else on the hill opposite. We can also see a local industrial estate.

KAT: How will you go about doing this?

KURT: It depends on how bold I am about it. The mechanic I take my car is in the industrial area. If he owns that building, he might be willing to put something up on the roof. Or I could go and do similar to what Pat is thinking about doing in his area, and that's letterboxing.

Kurt rejects the impossibility of getting connected and lists an array of strategies including bouncing signals from different points, modding existing antennas, letterboxing and targeting key people. Like Ben, Kurt exhibits extraordinary persistence in exploring alternative ways of getting connected and his solutions are reliant on a myriad of contingencies. Kurt cannot get connected on his own. But there is a chance he can with the help of others. Being 'bold', for instance, is a strategy for extending

DOI: 10.1057/9781137312532.0011

his network to strangers outside the group such as neighbours and his local mechanic.

These examples illustrate how the group is making a very different kind of technological infrastructure – a collaborative one. In many ways this means there are few limits as to how far it can extend across the city. However, getting connected to the network requires an intensity of engagement not ordinarily encompassed in the purchase and use of a commercial WiFi service. You do not simply *get* connected. You need to *give* time, tools, the roof of your house, electricity, investment of materials and money. You also need help – volunteers and family are vital in physically building and raising an antenna. And as it turns out, boldness, imagination and more than a little bit of hope are also critical.

The politics of the shared spectrum

Because WiFi operates on a shared public spectrum, community groups are not the sole users. ISPs, ham radio operators and a plethora of domestic wireless devices that broadcast electromagnetic signals within localised zones share this space. As discussed earlier, it is not uncommon for some signals to interfere or disrupt others. What is good noise to a WiFi maker might be something altogether more irritating to a different spectrum user. Haring writes about ham radio signals that occasionally 'strayed' into television and other radio frequencies causing distress and anxiety to those outside the group. 'Without realising it, a hobbyist chatting on the airwaves might produce a series of beeps and buzzes on the channel where his neighbour had hoped to find the night's baseball game on the radio' (2007:xiii). Several incidents of this ilk occurred during fieldwork. One involved the group being accused by a local ISP and also a ham radio operator of 'polluting' the communal spectrum with 'unnecessary interference' and 'digital rubbish'. WiFi makers were upset by this accusation and shortly afterwards posted a clarification on their website:

> Do commercial carriers have more rights to use Wireless LAN? *No, not at all!* This is because Wireless LAN designed for the standards 802.11a,b,g,n & s use a part of the radio spectrum which is free to be used by anyone, provided the radiated power is controlled. Called the 'Public Park Concept' it is relatively unregulated by the Australian Communications & Media Authority who allow all users the *same rights* regardless of who they

are – business, telecommunications carriers, government departments or private citizens. Beware of any misleading information to the contrary as there are some cowboy businesses who might try to have their customers and the public believe that they have special rights when it comes to the use of this radio spectrum and that other users create interference and are unauthorised or even illegally using it. (Emphasis in original)

This interaction points to the complex socio-political frameworks that structure the shared electromagnetic spectrum across the city. Echoing the multiple and shifting wireless noise present in and around my house, this larger digital landscape is not an empty, fixed or neutral space. Instead, we see a rigorously defined hierarchical system of power, occupied by many inhabitants who jostle for space on a daily basis. In another work, I have described the often inexplicable 'technological tantrums' that result from the introduction of WiFi into the existing domestic digital ecologies and the 'wireless workarounds' necessary to make possible everyday use (Jungnickel 2006). Turnball (2000) uses the term 'motley' to account for a diverse array of actors in heterogeneous science and technology networks in his study of malaria cures, medieval architects and aboriginal mapmakers. 'Feral' is a more local term for a similar concept. Feral is defined as a wild animal or plant that has escaped domestication and in Australia is used to account for the impact of introduced species into native habitats. Together with Bell (2008), I have written about commercial domestic models of WiFi as 'feral technology' that require an acute and flexible awareness of the fragile ecologies in which they reside and only become visible to users at points of attachment with specific devices, at interruptions or interferences with other domiciled artifacts. I suggest WiFi in this community context reflects that of a feral infrastructure. Ben and I encountered a technological system uncontained by conventional distribution systems of roads, pipes or cables. It evaded the constraints of traditional architectures of ownership of fences, doors and walls. However, upon erasing some actors, others become more visible, such as existing users of the shared spectrum who complained about WiFi makers invading their wireless territory. Here, WiFi plays the new (feral) intruder role in an established (native) technical landscape that has already been carved out by ISPs, traditionally larger and more dominant in the market place, and ham radio operators who have been around longer than community WiFi. Although the language of conflict appears extreme – 'polluting' and 'rubbish' – it is comparable, looking more broadly, to the way Australian

DOI: 10.1057/9781137312532.0011

customs with its island mentality deal with foreign fruit and animals; it attempts to identify, quarantine or eradicate them.

This chapter presents another example of the shift away from thinking of WiFi as a consumable and individualised plug-and-play technology to one that catalyses social engagement and sensitive awareness of radio emissions of other devices in the home as well as entanglements with actors in the broader socio-material and political landscape. Stumbling provides new ways of thinking about visible and invisible wireless noise that leaks out of the home, collaborative forms of social negotiation and the role of technological imaginings in the making and using of a new digital technology. STS literatures hold that seeing is closely tied to local, social and cultural ways of knowing. Here, the group demonstrates a form of connectivity in which knowing and not-knowing are purposely built into the culture of the network. Far from unsettling or destabilising it, the inability to produce fixed and certain representations operates to strengthen the group, expand the network and grow membership. The role of technological imaginaries in the group's making of culture is the topic of the next chapter.

DOI: 10.1057/9781137312532.0011

7

Mods and Modding

Abstract: *'Mods' are modifications that come about when things do not quite fit as a result of changing conditions. They represent an almost infinite combinability of ideas, materials and applications and demonstrate makers' aptitude for innovative responses. Drawing on examples I describe how makers mod not only technical materials but also find themselves tinkering with the broader technological landscape, social relations and stories as a means of dealing with socio-technical incoherence and instability. This chapter also discusses the Do-it-Together (DiT) nature of the network – the fact that it cannot be built alone. It requires the help of many others including families, partners, sisters, mothers, fathers and friends.*

Jungnickel, Katrina. *DiY WiFi: Re-imagining Connectivity*. Basingstoke: Palgrave Macmillan, 2014. DOI: 10.1057/9781137312532.0012.

'You don't need flash stuff, you just need sticky tape'

At a monthly meeting Ron invites Peter and Craig to update the group on an antenna they built and installed over the weekend on a domestic roof in the northern suburbs of the city. They tell stories using a series of photos projected on the wall from their seats in a semi-circle of 12 people. We see photos of the two men clutching bulging shopping bags in a local hardware shop, working with tools in a suburban backyard and squatting in the butted edges of iron sheeting on the house roof. Peter explains how they had to 'mod' (modify) their initial plans. The first mod came about when they were faced with a limited selection of tripod masts at the shop, which catalysed the decision to make their own. They purchased a nine-foot pole, some smaller lengths of piping, nuts and bolts. Returning to the house, they flattened the tubing for strength but found the bolts were too small, so they returned to the shop for new ones. Later, they encountered the extreme incline of the roof and more mods were needed to ensure the mast was strong enough to withstand the local 'gully winds' and resident birdlife renowned for sitting on antennas, yet could easily be brought down for repair and upgrades. They also had extra pressure to get the work done within a pre-agreed time to avoid inconveniencing the homeowner. But, due to the many mods involved, they had to leave before they could trouble-shoot an 'untrustworthy cable', so they made plans to return next weekend. The group claps when they are finished and Ron congratulates the men saying he loves their photos and tells them to post them on the website because they show 'how it is done from scratch' and 'ordinary blokes doing stuff'. Peter smiles and says: 'It shows you don't need flash stuff, you just need sticky tape'.

Peter and Craig are clearly experienced and were well prepared, yet things did not go to plan. They had to adapt and customise their approach, materials and tools. If WiFi makers were only interested in a finished antenna, a very different story might have unfolded. Perhaps then Peter and Craig might have waited until the antenna was actually working before telling the story. The very thing the antenna does, that it is built to do, is not the focus. Instead, Peter and Craig draw attention to the tangents, accidents and unexpected happenings along the way of making a new point in the network. They focus on a range of modifications, revealing in candid detail what did not work and why, and how they worked around each problem.

DOI: 10.1057/9781137312532.0012

Mods are modifications that come about when things do not quite fit as a result of changing conditions or available materials. They involve putting things together in new ways, combining assemblies of human and non-human actors into newly configured heterogeneous networks without flattening or erasing their individual shape. As a result, they are never finished. They materialise alternative interpretations and are suspended in the process of making. The fact that mods can be disassembled, and reassembled at anytime signals the flexible, infinite re-combinability of the socio-technical system and the value attributed to resourceful adaptability and ingenious practical knowledge.

BEN: A mod is fixing or modifying something for a better purpose.

SIMON: It's about finding the limits and potentials of the items. Some of the things I've got here I've modified to exceed what their specifications were when they were shipped. I guess you are trying to find something more by modifying it because it doesn't quite fit what you are doing. Or you are trying to do something slightly better, so you try and bring your model up to speed with what you believe you need to do.

TIM: [It's about] making it do something that it wasn't designed to do.

KERRY: It's mostly making stuff work... It's building things. It's part of the culture of taking things that weren't quite made to do what you want them to do and make them do something slightly different. It's the whole adaptive approach. 'Oh I wonder if I can just...' 'Oh, oh look at that isn't that interesting'. 'I'll make use of that'. So the original manufacturer wanted to do this really limited thing with it and people go, 'Hey you can do other cool stuff as well'.

RON: Sort of like changing the nature of something. If I buy a car I can pull the engine out, change the wheels and sell it to somebody else. Why can't I do that with a piece of technology?

There are many similarities between modding and tinkering. Tinkering is 'part of the inquiring approach to the material world' and 'includes scavenging, scrounging, tampering, adapting, fossicking, fixing' (Thomson 2007:6). Being able to tinker or mod reveals an ability to adapt to changing circumstances and unexpected happenings; skills highly regarded in the group. Modding demonstrates an aptitude to make things fit together, to push and extend things beyond what is expected. It springs not from smooth and exact results but from encounters with a range of contingent assemblies, tangents and variations. Although it takes many forms, modding ostensibly involves adjusting pre-purposed plans in line with material availability, characteristics of place, skill and time restraints. Modders see stuff differently; everyday materials such as

DOI: 10.1057/9781137312532.0012

sticky tape, off-the-shelf equipment and easily available materials from the local hardware shop take on new meanings. Modding takes the user into the black box.

On the surface, it appears that WiFi makers delight in the unpredictability and instability of their network. Mistakes, things that break and fail are regular features of talk, demos and photographs. Exposing how they deal with unexpected tangents and dead ends serves to reveal a technical curiosity and innovative ingenuity; a mastery over uncertainty. Other community groups such as ham radio operators share a similar tinkering technical culture. A unique feature of WiFi makers is the collective nature of this body of knowledge. Circulating modding experiences, in stories, images and objects, lies at the heart of these experimental encounters. Members learn from and build on these experiences, which deepens and strengthens their knowledge about WiFi in specific conditions and in turn grows the network.

Raising an antenna in suburban Australia

A few months into my year as a member of the WiFi group, I was offered a chance to help build and install a new wireless antenna. Dave, 22, a relatively new member at the time, posted a note on the group's public forum calling for 'a helping hand or two' to 'raise an antenna' on his parent's backyard shed in a suburb ten kilometres north east of the city centre. After responding to his post, he emailed me: 'Just to let you know that Sunday is going ahead. I am combining the AP [access point or antenna] raising with a BBQ with cooking starting at 11am'.

Most of the antennas in the community network, with a few exceptions,[1] were individually owned and located on house roofs, corrugated iron sheds or commercial buildings where members have family connections or employment. Although it is possible to be a member of the group without installing an antenna, it is a normative practice to at least attempt to raise one on or near your home. Dave recently graduated with a degree in engineering and computer systems. He had been working on contract for a variety of IT companies but was looking for full time work in the area of industrial automation. I met Dave at a monthly meeting and regularly saw him at public WiFi events such as LAN sessions. Reserved and soft-spoken, I initially thought Dave was a 'newbie' like me, and on several occasions I mistakenly introduced him to others

DOI: 10.1057/9781137312532.0012

until I discovered he knew much more about the group than I did. He had attended the group's first Open Day in 2000 and kept in touch with members via the website, attended meetings and 'officially' joined at the end of 2005 despite not being connected to the network. He explains:

> I wasn't really that well known. I was in the background a little bit. I used to go on the IRC [Internet Relay Chat] channel and used to know a few people from there…but I wasn't really involved in such a capacity as going along and meeting people and doing installs and things that I have been doing the past year.

Although the open visual culture of the group enabled Dave to participate from the outside, he did not really feel involved until he started contributing, which symbolises the importance of demonstrating interest. Raising an antenna at his house offered a way to connect to the network, technically and socially. Being a WiFi maker is not about simply using a technology. Members become part of a community, not just a consumer of a service. It is not a one-way relationship, but rather speaks of a multi-directional dialogic socio-technical engagement. The network is comprised of distributed people and antennas. No one person can make the network just as a single antenna does not work on its own. Smaller antennas connect together to form the larger network and individual participation and contribution are interconnected in a collective collaborative system. Participating in an antenna-raising event reveals how Dave became engaged not only with modding his antenna, but also modding social connections, found objects, meanings and stories about the experience.

Modding social connections

It is midday when I finally locate Dave's parents' house in a row of single storey brick bungalows set in neat, though very dry, garden edged lawns. I am later than planned as I briefly lost my way in the suburban sprawl. As I wheel my bike along the driveway I see two men standing around a smoky barbeque in front of an open and very full backyard shed. They spot me and the older man yells over his shoulder, 'Dave, one of your friends is here'. Looking past the two men and into the shed, I see Dave sitting on a garden chair with his laptop and surrounded by suburban detritus: old bikes, garden benches, paint tins, hammers and

DOI: 10.1057/9781137312532.0012

shovels, bits of wood, fishing rods, lawn mowers, hoses, insect repellent, garden poisons, coiled electrical cables, camping equipment and ladders. Two women swing on a sun seat at the edge of the house. After warmly welcoming me, Dave's mum and sister ask whether I want something to drink. I hear yelling inside the house and Dave's sister explains that 'the boys are playing games inside'. I ask if they are from the WiFi group. She shakes her head and explains it is Dave's younger brother and one of his friends who are also here to help.

Dave's parents' home presents a particular suburban vision of WiFi; a sprawling single storey brick bungalow, surrounded by neatly mowed yet drought ridden lawn, with a shed and small vegetable patch, a lazing cat on the driveway, the distant whine of a lawn mower, scorching sun, smoking barbeque and presence of extended family and friends. The use of residential homes as WiFi sites in the network is standard practice in the group. The network is made up of a series of interconnected antennas and members' homes provide crucial sites because they are free, easily accessible should something go wrong and provide ample space for modification and tinkering. Domestic infrastructures and relationships are pivotal in making this version of WiFi. Unlike other members who rent, Dave will not have to negotiate with a disgruntled landlord should something happen to the roof. Nor will he be required to dismantle the antenna, if and when he moves away from home. He does, however, need parental permission and support to build and install an antenna. In this case, he has successfully negotiated the roof of the backyard shed and access to a range of materials and tools. Furthermore, given the promise of a barbeque, he has secured the support of his family and friends.

Modding the 'barbie'

Dave introduces me to his brother-in-law and to his dad, who wearing an apron and flourishing a pair of tongs says, 'We're cooking Skippy'. Skippy, the name of a kangaroo, was a central character in a popular 1970s Australian children's TV program. Although kangaroo has been eaten in central Australia for many years, it is a relatively new source of meat in suburbia. Looking at the hotplate I see quite a lot of Skippy. Dave's Dad points out that in addition to kangaroo fillets, there are the usual chops, snags and onions. I lean my bike against the fence, peer around the corner of the house looking around for the antenna, the

DOI: 10.1057/9781137312532.0012

reason why I am at the barbeque, but instead I see a large ping-pong table covered in a floral plastic tablecloth and set with twelve places. I am relieved that I am not late but I start to worry that I should have told them I am a vegetarian.

As discussed in Chapter 4, barbeques are regular features of WiFi events such as summer meetings in the public school quadrangle, at large installation events and Open Days where antenna shoot-outs and product demos are set up next to snag packed hotplates. Barbeques also feature prominently on the group's website. An edited photo of a snag skewered antenna signals the normative intertwining of WiFi and barbeques. I initially found the combination of wireless technology and scorched meat surprising, yet the WiFi barbeque is less about the actual food on offer and more about the socio-technical interactions it makes possible. Barbeques fit with WiFi because both are socially constructed. Unlike hackers who are 'not joiners' (Wark 2004) and tend to be engaged in their pursuit in the isolated privacy of their bedrooms or offices, community WiFi networks hinge upon sociality. Antennas need to 'see' another to make a connection and are often disconnected due to a myriad of factors; therefore members rely upon social connections to ensure technical ones are made and constantly maintained.

Yet, here was a table set for an intimate family meal (Figure 7.1). This was not the basic affair of non-descript sausages, onions, bread and tomato sauce, prepared and eaten while standing as per group meetings. The placemats, special meat and presence of extended family suddenly made me feel that I should have pre-warned Dave about my dietary requirements. I had not thought my eating habits were particularly relevant to my study about WiFi. I had not expected the barbeque to play such a pivotal role in the home context because in the local school playground it operated to domesticate public space, to render it intimate and social. The home was an already familiar domesticated environment. Yet, even here the barbeque played a key role to link disparate people, objects and places together.

Dave's dad yells out that 'Skippy is ready' and brings a tray of sizzling meat to the table. It is shortly accompanied by a range of salads: a yellow curry rice salad, green salad, half boiled eggs with curry powder, tomato sauce and a bag of white bread. I take a seat and ask Dave if anyone else is coming. He shrugs. He thinks Simon and maybe Craig and some others but 'it depends who is awake'. Everyone agrees that lunch should start and if anyone else comes, they can join in. The ping-pong table

DOI: 10.1057/9781137312532.0012

FIGURE 7.1   *The family barbie of 'Skippy' before raising a new antenna*

quickly fills with Dave, his friend and younger brother, his sister and her husband, his parents and me. Almost an hour later, Simon walks around the side of the house and sits down at the table. Dave greets him warmly, introduces him to his family and passes him a plate of 'Skippy'.

Although Dave graciously accepted my 'dogsbody' assistance for carrying and holding things, he was clearly relieved to see Simon arrive. Simon, 22, was an IT student at the local TAFE (technical college) and a technical support volunteer for a local primary school. While he had less money to invest in the network than others, he contributed his skills and time and was a regular volunteer at new antenna installations. Simon explained his interest and experience came from 'playing around with computers' when he was 12 and he had been a 'long-term' WiFi member since 2002.

As will become evident, making plans, gathering equipment, preparing sites and securing helpers for an antenna-raising can take months. Barbeques consolidate these heterogeneous entities, bringing them and WiFi down to earth. They also fortify the importance of social connections. Barbeques transform WiFi makers into family and family members into helpers. The assumption that I am Dave's friend when I

DOI: 10.1057/9781137312532.0012

arrived, as opposed to just a WiFi member, signals the importance of social networks in making wireless technology in this context. Similarly, the warm welcome to Simon, despite his lateness, points to an intimacy beyond that of a casual helper. Barbeques serve to further embed WiFi in the domestic footprint, representing Dave's family's approval and support and operating as a method for enrolling and persuading volunteers to donate their personal time on a sunny Sunday afternoon.

Modding found objects

Dave crouches on a rectangle of lawn between the vegetable patch at one end and a shed at the other (Figure 7.2). A dry water feature, a 'whiz bin' (recycling bin), empty birdbath and colourful pot plants line the corrugated iron fence. CDs dangling on string in the fruit trees twist and glitter in an attempt to keep hungry cockatoo's at bay. With a silicon gun in one hand Dave peers close, touches and occasionally stands back to look at a steel tubular pole that lies suspended between a canvas camping stool and a wooden garden chair. A coil of blue cable lies on the grass at one end and a wireless dish connected to an omni antenna and a white box is U-bolted to the pole at the other. A house brick keeps the dish on an angle. Various tools, sticky tape and sunscreen lie scattered nearby.

The pole came from Mildura, a large country town on the border of South Australia, about six hours drive from Adelaide. Dave explains he 'heard about a pole' at his aunt's property, which signals his extended family's awareness of, and interest in, his project. He was not sure if it was possible to transport such an object, but his 'resourceful' aunt contacted the local

FIGURE 7.2   *Dave tinkers on his antenna in the backyard*

DOI: 10.1057/9781137312532.0012

shipping company, which turned out to be 'a man in a shed' who apparently was not surprised by the request. The man was used to people 'turning up with ten sheep and wanting them sent somewhere, so a pole was nothing'. It cost Dave $20 to have it sent to Port Adelaide, about 20 kilometres west of his house. He borrowed his Dad's ute (utility vehicle) to pick it up.

Finding and hearing about things is in keeping with modding and tinkering. Although Dave purchased several new items such as cabling and U-bolts that attach the box to the pole, most materials were gathered from his home, shed and personal connections. In addition to the pole, Dave found the box that houses his computer equipment – an old biscuit tin – in the kitchen cupboard. To counter the effects of the sun, Dave carefully painted the entire box with several coats of light coloured hardwearing and reflective house paint that he found in the shed. He was using a 5.1-GHz (gigahertz) antenna built by a WiFi member in Tasmania who had sent it to the group as a 'tester'. Although, the antenna is an assembly of found, bought and re-appropriated materials, looking closely at the contents of the biscuit box reveals an attention to detail. The motherboards are painstakingly mounted to provide access and ventilation. An ethernet cable is neatly coiled and tied. Drilled and filed holes in the top of the box for ventilation have been insect-proofed with neat squares of flyscreen. Two other holes firmly hold the large U-bolt with which it will be secured to the pole. The wireless pigtail has been inserted into another neatly drilled hole and sealed into place with clear silicon putty. The combination of ad-hoc materials with sophisticated technical knowledge, attention to detail and improvised methods gives Dave's antenna a unique character (Figure 7.3). It also signals that mods and modding are not incommensurate to professional skills and outputs.

FIGURE 7.3   *Inside and outside the antenna housed in a biscuit tin*

DOI: 10.1057/9781137312532.0012

Modding the mod

Dave and Simon climb onto the roof to measure the distances they will need for the guy wires. The pole will be bolted to a bracket in the centre of the shed roof and these wires will secure it at four pre-drilled and cemented hooks in the aluminium sheeting. Measuring five metres across the yard, Dave has me stand at one end on the curling wire. He uses the combination of my body weight and the yard to measure consecutive lengths. While we do this Simon tells us about another antenna he helped to install that was held with five guy wires set three metres apart. This made it very secure; however because they need to do some changes, they have to work out how to access it. Triggered by this conversation, Simon reminds Dave to test the remote switch before we install the box on the mast.

WiFi makers' awareness of the instability of the network shapes the design of new antennas. Installing a remote switch is an important feature of Dave's new antenna because, in theory, it means he should not have to take the entire setup down if there are problems. Intermittent connectivity is part of every WiFi network; therefore, it holds that Dave builds in alternate ways of accessing the antenna. Dave is building the node to withstand harsh weather conditions, as well as local fauna and flora. He has chosen the highest point available to him – the roof of the shed. It is being built to last, but it is also designed to change. Every antenna in the WiFi network is hand-made, either entirely or partially, using a range of available materials, improvised methods and personal skills. Antennas in the WiFi network are made differently from the beginning and changes are more distinct and abrupt, often catalysed by a constellation of unknown actors. As a result, Dave needs to build in ways to mod the mod.

We hold different ends of the pole and slide it through the garden chairs which have been operating as supports. Simon takes the lead and hoists it into a vertical position, leaning it against the edge of the shed. Dave plugs the electricity and the ethernet cables into the shed and does preliminary tests to see if it is working. At this point, more people arrive to participate in the raising of the antenna. Dave's dad, his brother and friend, Josh, come into the yard to help. Dave's father holds the ladder while Josh and Simon climb onto the roof. Dave and I help hoist the antenna up over the edge of the shed. Together they hold it in place in the centre seam of the roof. Dave untangles the guy wires and pulls

DOI: 10.1057/9781137312532.0012

one to the edge of the roof where he has earlier drilled in hooks. With his dad's help he secures the wires to the four points on the roof. Upon direction, I pass various tools that lie strewn on the grass and garden furniture. Dave's mum brings a tray of cold drinks from the kitchen and offers sunscreen to combat the fierce late afternoon sun. Even my camera is having trouble adjusting to the glare from the reflection on the galvanised shed roof.

Building Dave's antenna involved Simon, Dave and me. Raising it involved a much larger contingent. Dave, Simon, Dave's brother and his friend Josh, his father and mother, Dave's sister and brother-in-law, Simon and myself as well as Craig on the phone were all involved in varying degrees (Figure 7.4). After the steel pole is secured to the top of the shed with the guy wires Simon returns to his laptop on the ping-pong

FIGURE 7.4   *Raising a community WiFi antenna is a collaborative effort*

DOI: 10.1057/9781137312532.0012

table. He talks on his mobile to Craig who is at his home scanning for Dave's new antenna. Craig cannot 'see' Dave. Dave is on his laptop in the shed. He keeps rebooting the software and scanning for Craig's antenna. Dave is worried that maybe the antenna does not work after all.

It was past six o'clock in the evening and I had spent over six hours with Dave and the others in his backyard. Although the antenna was now secured in place on top of the backyard shed, it was still not part of the network because it had not been 'seen' by another antenna. Without a connection, an antenna is just an antenna, an individual point, isolated from the network. Nothing happened for a further half an hour, when suddenly Dave burst out of the shed and grabbed my shoulder shaking it in excitement, saying 'He sees me, he sees me'. At this point, Dave's antenna became part of the network.

Modding the story

Shortly after the installation, Dave published his experience on the Air-Stream forum. He described in detail the events of installation and did not revise the narrative to hide the mistakes or awkward aspects that resulted in less than perfect results. His representation featured non-standard materials and improvised techniques employed in his setup, including the fact that at the end of the day, the antenna did not maintain the connection. Despite all of his preparation and work, Dave described in candid detail his inability to get connected. Over subsequent weeks, he published more posts documenting two further attempts, including a late night session in strong wind with fewer helpers. I also overheard him telling it to a cluster of members at meetings. Far from undermining his abilities, the dramatic story served to draw attention to his ability to deal with unexpected challenges.

Orr (1996) calls these 'war stories'. Writing about people who repair photocopiers, he notes how much of their work is talk. 'A coherent diagnostic narrative constitutes a technician's mastery of the problematic situation' (2). Here, the telling of stories is also a vehicle for technical mastery, a way of diagnosing and addressing problems and demonstrating ability. In the context of WiFi, stories of mods enable the maker to make sense of the instability of the technology, to hold and harness what might otherwise seem to be a disparate and uncontrollable technology. It also provides the means for members to form bonds in the absence of conventional technical connections. When asked why he narrated such a candid account, he explained:

> Probably the main reason for documenting things like that and reading those pictures would be to show someone who's not quite sure. They want to do a similar thing or they want to copy it. You show them for example the inside of the box, like the two boards sitting in there. That's an idea that could be shown to someone else. They might have an idea of 'Ok I want to use two of these, but how can I do it?' So, 'Here I've done it before' and 'Have a look'.

Dave's account shows that there is no one way of making WiFi, there are many ways. The collective Do-it-Together (DiT) culture of the group expands to encompass multiple descriptions of how WiFi is made. It does not matter if these descriptions are of working or non-working devices, of mistakes or failures. The WiFi group eschews a triumphal point of expertise. Members promulgate a flattened topography of knowledge in which anyone can contribute. From the initial invite through to the raising of the antenna, Dave gave no indication that revealing, or of me documenting, his modding was in any way problematic or cause for anxiety. Instead of tidying up loose ends and narrowing or entirely closing down alternatives by cementing ideas in place, WiFi makers like Dave, Craig and Peter assemble and constantly reassemble shifting, malleable and complex materials, places and roles in the everyday making of WiFi. At no point can one single or definitive version of the network be fixed or known, rather, it is only temporarily stabilised in order to allow members to flexibly respond to the relentless uncertainty of their local context. Telling stories and representing their responses to difficult situations and conditions create coherence across incoherency. As a result, there is no dominant or linear path to innovation. Members are expected to make their own stories of their own WiFi experiences as they go along, to apply resourceful ingenuity to constantly emerging uncertainties and contingencies and most importantly to share them with others.

Modding the meaning of connectivity

A significant part of WiFi work for makers lies not only in modding sites, materials and social networks but also meanings about wireless technology. According to Jan, who often talked about the ontological differences between 'makeaholics' and 'shopaholics', people are so used to expressing themselves as consumers, especially in terms of the internet, it makes the concept of community WiFi networking difficult to explain. People tend to think of connecting in terms of a transaction, which in turn shapes their means of expression. They pay for a service and feel they must get something in return. As a result, WiFi makers often find themselves in the role of re-educating people.

DAN: ... it's a foreign concept ... people are so used to the internet they are very keen to adopt that description of it ... I think I read a statistic that says ten per cent make and 90 per cent consume. So it's like the majority of people don't. They just go on there and look it up. Whereas Air-Stream is about creating your own website or put images on there or you'll write something and that's why we have a website completely full of information. You create things yourself and put them on the network. You actively participate in it. You are not just a consumer. Which is a key difference I think to the internet.

RON: We see people who think they are going to get cheap internet or are becoming a member because they are going to get a service for a fee: 'Oh its 50 bucks a year and I'm going to get something from this'. And they don't understand that it's actually an association and you only get out of it what you put into it. But you soon see people willing to go, 'Oh I got all this information', 'I've got this and I've got that', 'I'll come out and help you set up' and 'I've got this extra part I can lend you' and are willing to share and aren't guarded about it.

Air-Stream's approach to WiFi is not instinctive to many; especially those who expect to pay and receive something for the transaction. Dan and Ron emphasise how the network does not work unless people get involved and contribute and this requires a paradigm shift for those who subscribe to consumer models. Although people know how to connect via conventional means, they do not know how it could be otherwise and what role they might play in re-making it. In line with this, the multi-faceted visual culture, mods and hands-on approach of the group operated to render visible new points of attachment in the hope that one of them sticks.

Another meaning of WiFi that the group attempted to challenge was the solitary stereotype of tinkering. In their study of robot builders and professional software developers, Kleif and Faulkner (2003) found tinkering to be a solitary affair. They write about how 'few of the men admitted to any enjoyment of or competence in working closely with people' (301). Similarly, Turkle's (1995) computer programmers were individual tinkerers and Thomson's (2002, 2007) research often depicts the shed as a refuge from domesticity and the world at large, inside of

DOI: 10.1057/9781137312532.0012

which is depicted a lone man engrossed in a task at hand. In contrast, WiFi tinkering hinges on social connections. It is best described as collaborative and tasks are group-oriented as evidenced by the meetings in Chapter 4 and Dave's antenna raising. Although it is possible to stumble alone, the group's website recommends members take 'a buddy' onto rooftops to hold antennas and ladders in order to prevent accidents. Working with others is encouraged and in some cases actually necessary for a job to be accomplished. Ironically, what these examples point to is the inappropriateness of the term 'Do-it-Yourself' (DiY) in relation to the practices of backyard technologists. WiFi makers are clearly not individuals who only subscribe to a DiY approach. They Do-it-Together.

While group practices contradicted some conventional understandings, other aspects of tinkering were harder to shift.

SIMON: When I was in year seven I think I got my first own computer that was mine and stuff to break and fix. And ever since then I've been intrigued by everything that is computers, any aspect of it. I've just wanted to get into it, to play with it and programme it or create it and know more and more about it…It was just who I was…It was just what I did.

KERRY: It's the whole adaptive approach. It's what I've always done.

DAZ: I broke too many things when I was young. I remember when our washing machine broke and a whole section of it was pulled out and replaced and I got to play with the broken bit. So when things used to break, like a radio or something, I used to want to open it up and see how it worked, things like that. I was always the person who gets asked to fix something, like a VCR or something, like 'What wire goes where?'

According to these makers, tinkering is not something that can be turned on or off. Making-do, as I discuss further in Chapter 8, has also been traditionally regarded as resolutely masculine (Thomson 2002; Jackson 2006; Bollen et al. 2008). Jackson notes how 'many writers have referred to 'rough and ready' local designs with a certain measure of pride, as if this characteristic in some way attested to their masculinity' (2006:253). The fact that WiFi makers describe it as a 'natural' practice from a young age fits with what has been written about tinkering by Kleif and Faulkner: '[B]oys are more likely than girls to be socialised into hands-on tinkering with mechanical devices' (2003:297). Some WiFi makers' experiences echoed this perspective:

DAVE: I used to try and incorporate my sister into it a little bit. She wasn't really interested in how I looked at things. I would make games so she would be involved. Like we'd play a library game where we'd lend each other our books and I would write a computer program.

KAT: And your sister? You said she wasn't so keen on this?

DAVE: She just cared about the books and I just cared about writing the program.

The WiFi network appears open to everyone, and women are clearly entwined in how it is made, yet the traditional masculine cultures of tinkering provides a potential explanation as to why WiFi appears to 'stick' to men (Faulkner 2000, 2006). In a brochure produced for the engineering industry to assist in the recruitment of women, Faulkner argues that the traditional hands-on identity of engineering does not reflect contemporary practice and moreover, serves to alienate women: 'Arguably, the "nuts and bolts" identity is a comfortably "masculine" one for many men, but it can serve to exclude other engineers because it does not to capture the diversity of engineering work' (2006:2). She points out that this has the effect of attracting only a certain type of men and therefore alienates more than just women. The way Australian men 'take' to WiFi is also reminiscent of how Trinidadians took to the internet (Miller and Slater 2000). Miller and Slater describe how Trinidadian's saw the internet as being about chatting and because that is what Trinidadians did, they developed a 'natural affinity' to it and thought about the internet as 'naturally Trinidadian' (2). Despite the work being done by the group to widen its appeal, the masculine comfort and pleasure that derives from tinkering and making-do with WiFi complicates the group's task of attracting new potential makers, such as women and men who do not embrace this version of tinkering and making-do technology culture.

This chapter has discussed the DiT nature of the network through mods and modding. Mods reveal quick, ingenious and resourceful responses to deadlines imposed by daylight, the home-owner's schedule, skills and financial restraints. They enfold materials 'at-hand' and incorporate improvised methods. Mods and modding are essential to experimentation. They emerge from a deep understanding of a constellation of socio-technical properties. Teasing them apart reveals a process of learning through trial and error. WiFi makers appear to delight in the unpredictability and instability of their network. That mistakes, things that break and failure are regular features of how members make WiFi accounts for why they are not embarrassed about their exposure, such as when Dave candidly revealed his process of problem solving. Exposing the many tangents and dead ends served to reveal his technical curiosity

DOI: 10.1057/9781137312532.0012

and innovative ingenuity. In this way, members build on each other's experiences, which deepens and strengthens their knowledge about WiFi in specific conditions.

Mods produce alternative interpretations. They are critical to the production of wireless knowledge. Just like Dave's unique interpretation, there is no dominant, linear or central path to travel. Members are expected to make up their own stories as they go along, to apply their own resourceful ingenuity to constantly emerging uncertainties and contingencies and most importantly to share them with others and build on each other's innovations. Instead of tidying up loose ends and narrowing or entirely closing down alternatives by cementing ideas in place, WiFi makers like Dave, Craig and Peter assemble and constantly reassemble shifting, malleable and complex materials, places and roles in the everyday making of WiFi. At no point can one single or definitive version of the network be fixed or known, rather, it is only temporarily stabilised in order to allow members to flexibly respond to the relentless uncertainty of their local context.

Central to this practice is recognition of diversity and individuality as strengths in the group. Here, the possibility of alternative approaches signals opportunities, not disaster which means the culture that surrounds this Australian version of WiFi encourages people to participate, and contribute, not just consume and use it. The group is not just about providing wireless internet access, it is a platform for experimentation. WiFi makers present a way to re-articulate the way we think about, communicate and understand new technologies, and their role in our lives. The group's mods re-script innovation; not as a hidden process, nor within a spectrum of success or failure, but in relation to being collaborative and constant. This goes some way to explain why the community network continues to expand despite the growth of cheaper and more easily accessible commercial internet provision.

Note

1   Some antennas are collectively owned by the group and located in a central location such as the rooftop of a local hospital, which was negotiated by a member who worked there. Group members donate materials, money and time to build and install these antennas.

DOI: 10.1057/9781137312532.0012

8
Homebrew High-Tech

Abstract: *This chapter explores the seemingly contradictory intersection of homebrew and high-tech. Drawing on encounters between local ISPs and WiFi makers, I argue that this conflation signals a distinctive cultural way of imagining and making a version of wireless broadband highly localised to the suburban backyards of Adelaide. Building on the previous chapter, I examine the role of 'making-do': a distinctly Australian version of modding interlocked with the peculiarities of the local landscape, weather and colonial history. Sticky tape is also reviewed as a mundane tool and symbol of a way of working. Both concepts represent unique ways of re-imagining how innovation and inventiveness happen in the suburbs of Australia.*

Jungnickel, Katrina. *DiY WiFi: Re-imagining Connectivity*. Basingstoke: Palgrave Macmillan, 2014.
DOI: 10.1057/9781137312532.0013.

'Making-do' – an Australian approach to inventiveness

> Our national knack for invention and innovation, for making do, lives on in the shed. The 'she'll be right' attitude may be denigrated as the blight of Australian industry, but it thrives in the country's backyards. (Thomson 2002b:3)

Because WiFi constantly breaks and requires mending using a combination of improvised methods and available materials, it fits with the idea and the practice of making-do. Making-do is a distinct approach to technological innovation and adaptation borne of intractable places and conditions. Yet, as Thomson (2004, 2007) argues, making-do in Australia is perceived both as a source of national pride and embarrassment. Belittled in a national public 'front yard' context, it nevertheless holds great significance and value in everyday 'backyard' practices. One way to make sense of this tension is by exploring historical accounts. As outlined in Chapter 3, Australian identity remains influenced by outback imagery despite the fact that the majority of people reside in coastal capital cities. Given the reality of suburban living, I start by exploring how this enduring bush legacy shapes attitudes to technology.

Making-do emerged from the peculiarities of harsh bush conditions, economic struggles and limited materials of Australia's colonial past – a survival technique that fused resourceful local knowledge, ready to hand materials and hands-on skill. Although white settlers brought tools, building materials and established ways of working in Australia, they had not planned on a fundamental difference: the landscape and ecosystem. Rarely did technology produced in other countries simply work in Australia. Imported tools inevitably required adaptation due to a combination of sharp differences in topography, magnified scale of use and drastic shortage of labour, which gave rise to the practice of making-do with what you had.

> With transport slow and distances from cities great, an ability to solve any number of small engineering or manufacturing problems was necessary for a farmer's survival. A broken plough could not be repaired by a quick phone call or a part trucked up from the city overnight. The problem had to be fixed through ingenuity and resourcefulness. (Thomson 2002b:8)

Although South Australia became known as the '"hearth" of agricultural innovation in Australia' as a result of many successful technological

DOI: 10.1057/9781137312532.0013

adaptations, Birmingham et al. (1979) note how the *Australian Settlers Handbook*, produced for early English immigrants, failed to recognise the legitimacy of these inventions:

> James Atkinson in 1826 recommended the immigrant to bring a swing plough and the irontines for a harrow to be made up from a forked log of wood...Little had changed by 1861, when the *Australian Settlers handbook* recommended much of the same list: eighteen years after the invention of the stripper, the prospective settlers of New South Wales were told to bring sickles and reaping hooks, and sieves for winnowing in the breeze. (17)

The fact that local technologies did not make it into the handbook suggests they were not considered of significant value to replace long established British agricultural knowledge. Despite widespread success, local inventions were seen to be of questionable integrity due to their adapted, cobbled together and re-purposed characteristics. They had been tinkered with. The very essence of what made these technologies work successfully in challenging conditions became grounds for their dismissal, which points to tensions inherent in the country's penal heritage with its attendant class implications.

Haring (2007) suggests that socio-economic judgements play a role in the trivialisation, or in this case total rejection and subsequent erasure, of tinkered technologies. In identifying the division between people who learnt technical knowledge in classrooms and others who gained experience on shop floors or on their own, she writes of the distinction between an 'association of study with the wealthy and of tinkering with the working class' (90). The tension inherent in the relationship between tinkering and sophisticated innovation signals that hands-on does not naturally intersect with high-tech. Another way to view this is through what Henderson calls the 'aura of high tech', which she argues shapes how engineers work and represent themselves (196). She illustrates how a high technology, in this case computer assisted design software (CAD), has more status than one that is considered low, such as hand drawing skills. This, she argues, goes some way to explain why engineers use CAD despite the fact they know it can never fully capture the tacit or personal knowledge of the 'pencil and paper' doing of engineering (197). By comparison, I was intrigued by this comment by a WiFi maker in a monthly meeting: 'It's a homebrew telecommunications network'. The idea of *homebrew high-tech* highlights a fusion of seemingly incommensurate entities. While the former is nominally the

DOI: 10.1057/9781137312532.0013

reserve of hobbyists and amateurs in their spare time, the latter is most often associated with official well-established commercial organisations. In the Australian context, homebrew typically refers to the brewing of beer for personal consumption on a small domestic scale. As the name suggests, the home is central to its practice. This definition of the term 'homebrew', however, suggests something else. More is going on here than a domestic leisure or labour designation. Air-Stream's version of homebrew is not confined to the edges of the home or the sole pursuit of a lone individual in the domestic sphere. It extends out into the urban landscape and involves a broad participatory community. The chapter further explores these contradictions. WiFi is a fragile, often unpredictable and complex technology that predicates meticulous accuracy, sophisticated technical skills, the understanding of complex visual schema and specialised materials. It is also a technology that WiFi makers put together in their backyards, using customised combinations of purchased, home-made, found and even partially broken materials. How do WiFi makers negotiate the intersection of precision and tinkering? How do they straddle the DiY and professional divide? How does homebrew fit with high-tech? How, if at all, does it shape who can and cannot participate?

Building (and maintaining) reputation

I cycle to Simon's home on a hot Sunday afternoon in the north west of the city. Simon lives in a corrugated iron shed, the size of a double garage, at the back of his parents' house. Inside, it looks more like a messy home office than a garage or a bedroom. Tucked away along the left side wall is a small fold-out bed, but like the many differently sized desks, a well-used lounge, wall units and chairs scattered around the room, it is covered in empty chip packs, plastic coke bottles, piles of CDs, keyboards, books, spare monitors and cables. Stuff, as Thomson (2002a, 2002b, 2007) in his many shed studies has pointed out, is fundamental to resourceful DiY practice. During a brief tour, Simon points out the clear side panel on his external hard drive that reveals the meticulous presentation of the internal components.

> It's just so I can still see what's in there. When it's all boarded with steel sides you can't see what's in it. So having a clear case means I can see what's running and I can see what I've got in it.

DOI: 10.1057/9781137312532.0013

Simon clears a chair of debris for me and we talk while he sits at his desktop computer flicking through photos of recent antennas he has helped to build and install. He always carries a camera with him, takes photos and short video clips, often posting them online and sending copies to people in them. In one close up shot he points to out details of cables that are folded 'just right' and 'clipped tight', and explains how everything is clamped together 'all nice and neat'. The contradictions to his living space are rendered all the more striking in contrast. I ask why this was important:

> Because it looks professional. Especially with the Air Stream stuff because it's all voluntary. We are not getting paid, and a lot of people see it, [so] we try to make everything look as professional as possible so when they see it, 'Oh it must be done by a business' and then they realise it's the Air Stream equipment. 'Oh wow that's impressive'.

Being and looking professional is deemed important by WiFi makers in creating a space in the public imagination that equates professionalism with commercial business. Reg had similar convictions. He showed me an antenna installed at his place of work. After climbing through the service shaft on the fifteenth floor top of a state government health centre we emerged on the rooftop. Given the view that stretched all the way to the Adelaide Hills I could see what Reg meant by this being a 'good site'. It had an uninterrupted line-of-sight over much of the city. I noticed two identical antennas fixed on the wall. Everything from the type of pole, brand of box and dish, to the way the cables were curled and secured was the same. I asked if both were made by Air-Stream. Reg smiled and shook his head knowingly. One was owned by a local ISP. He asked me to guess which. I couldn't. He asked me to guess which one was there first. It wasn't the ISP's. In an interview later, Ron explained more:

> They are using the same equipment. And I know for a fact because that equipment is owned by [name of ISP]. A lot of his knowledge came from me, and that's why he's using the same boxes as me.

These examples illustrate a version of tinkering that does not produce amateur work. In the first, Simon reveals the desire for highly skilled expert results comparable to commercial provision. In the second, the fact that a commercial ISP copied the work of the community group is testament to the proficiency and skill of WiFi makers. Of course, as is clear from previous chapters, WiFi makers also delight in producing far less complete devices and installations.

DOI: 10.1057/9781137312532.0013

KERRY: A lot of the other guys are really keen on making and putting stuff together and if it works its fine or if its bodgy and there's tape hanging out the side so what, it's no big deal. It works, right, that's the most important thing.

This account suggests that it does not matter what the object looks like, provided it does the job. Tinkering, like other activities, is also about contradictory impulses, co-located forms of knowledge and a complex array of representations. WiFi makers are not trained in neat, linear trajectories or expected to conform to identical versions of the technology. Instead, individuality and resourcefulness is encouraged.

Haring has argued that ham radio operators 'proudly adopted the label "amateur" to stake out a certain independence' and 'never existed in a tidy dichotomy with a particular group of professionals' (2007:88). WiFi makers similarly occupy a blurred space between hobbyists and professionals. These categories are not easily defined and the group does not demand these distinctions are made. The network is built for modding, which makes it particularly attractive to a wide range of people. Reg, for instance, also works full time as an IT specialist:

It's like I can test it and play around with it and do whatever and not have someone say, 'Hey you're not meant to do that', or 'You shouldn't have that case open and those wires there'. It's a lot more relaxed. And if something breaks then you and just other people around the area are going to be affected. Obviously they might be a bit disappointed that you've played with something and broken it but...

As discussed, many makers work across the IT spectrum, some for the largest telecommunication organisation in the country and local ISPs while others ran their own businesses in IT related industries, consulted or volunteered for local schools or charities. This makes the division between hobbyists and professionals difficult to sustain. Kleif and Faulkner discuss the 'fuzzy boundary' between professionals and hobbyists when they write: 'Just as many of the robot builders had jobs working with technology, so many of the software developers had technological pursuits in their out-of-work lives – they did home maintenance and remodelling, they had leisure pursuits that used the latest gadgets, and/or they read science fiction' (2007:301–302). While community WiFi work follows a similar arc, much tinkering can be spontaneous, taking place in the middle of the night, during weekends and holidays or 'on the fly'. It is flexible and accommodating to other commitments and pressures in everyday life. The group's visual culture matches this practice as Simon explains his use of the IRC:

DOI: 10.1057/9781137312532.0013

SIMON: It's just the chat programme we use. You can have a lot of people on a channel. It's well set out so you can easily see who's saying what.

KAT: How often are you on it?

S: Oh, I'm on it twenty four seven.

K: So it is open now?

S: Yep. I just jump in and see who's around. It is usually the first point of call for people. Everyone checks it...you can type something in the IRC server and it will sit there and when they get there they can have a look or someone else can help out...It just adds the social aspect to the group because everyone is in there and they are just talking about anything they want at any stage. I guess it's like if you're all working in the same building and you go have lunch and you talk about stuff.

IRC is specifically built for synchronous group conferencing and runs continually which means traces of conversations are never erased, nor are they private (unless specifically intended) from the rest of the group. Makers can catch up on things they might have missed and learn from others' interactions. Being able to pick and leave off where they started suits those who are committed to other activities. The use of IRC supports the fact that tinkering does not happen in isolated blocks of time. It importantly captures a visual history of the DiT culture of the group. It is also not limited to Air-Stream members. Connections have been made to other community WiFi groups, such as one in Perth, to enable a wider discussion of problems and sharing of knowledge.

Backyard technology in a commercial world

Because they operate on a shared spectrum, WiFi networks are shaped by a group's relationship with, approach to and understanding of local ISPs. Makers foster a distinct engagement with larger socio-economic and political forces by constructing and representing WiFi as an open source technology. In Air-Stream's case, far from viewing the world as a competitive arena or a stable and impenetrable system in which passive consumers await the provision of, and pay for, services from large commercial models, participation and contribution in the infrastructure itself is encouraged. Air-Stream's approach to WiFi is very much a part of their larger homebrew high-tech ethos. While other community WiFi groups compete on a commercial scale and view success in terms of decreasing their opponent's market share Air-Stream foster a more

DOI: 10.1057/9781137312532.0013

symbiotic relationship that is in keeping with what Haring has written about ham operators in the 1950s: 'Electronics manufacturers especially, but other technical companies as well, sought to increase the number of hams on the payroll' (2007:83). The number of IT professionals who are also WiFi makers is a contemporary model of this. Ron believes that groups like Air-Stream would not exist without ISPs, and correspondingly, ISPs draw core knowledge from community groups:

> I think there is a really important place for commercial telecommunications. It just wouldn't happen at the scale that we all benefit from if it wasn't for commercial wheels driving that whole thing... I thoroughly believe that no business would be using wireless LAN for connecting business and providing internet access and hotspots if it wasn't for community people using it in the first place. It was designed for extending the local network inside your business. It was never designed for long haul links and it's only been enthusiasts who originally built their own high bandwidth antennas and set up long haul links. It's that open source mentality in that you share information and people share information back with you. You learn more. You gain more.

Ron's comments signal a DiT relationship between commercial and non-profit WiFi groups. Wireless technology was never designed for long haul links, yet because WiFi makers experiment and build their own homebrew versions it has enabled new uses and applications to flourish. Concurrently, access to high-tech equipment and the shared spectrum has enabled amateurs to operate. Drawing on open source ideology Ron explains how information collaboratively flows. The more you give the more you get in return. Knowledge is assembled and re-assembled according to who is involved and the nature of their needs. It is a conversation between many differently constituted and conventionally divergent bodies, producing a landscape where multiple versions of wireless technology are possible.

An illustrative example in which this kind of collaborative practice was enacted was when local IT businesses enrolled WiFi makers to experiment and test new devices. Simon was given one of two routers 'to play around' with, which involved arranging and rearranging materials into different configurations.

> I opened up the box and pulled it to bits straight away. I had heard from someone that there was a mini PCi card in it so I worked out how to pull it apart... and got it working in my laptop so that we can play around a bit. I've have left it open so we can change things again.

DOI: 10.1057/9781137312532.0013

These devices are not prototypes but finished working devices. The high-tech designers already know what their product can do within specific business and domestic parameters. They wanted Simon to explore the ways in which it might be used outside these boundaries. They were interested in *what else* it could do. The fact that Simon left the box open in order to continue to tinker and 'to show people what is in there' reflects the group's modding practice. The result of his experimentation was a review published on the group website. Asked why he thought commercial businesses do this, he explains:

> They know we are a fairly decent WiFi group. Our website ranks hugely on Google because we've got all the documentation, the WiKi and stuff. You type in the type of card and Air-Stream will be in the top three [results]. It's just been around for so long. I think we've had a couple of other pre-release and cheaper things but they are starting to realise that if they give stuff to us, people listen to us, 'Hey this is a good product. Go and buy it'.

The group's website provides a means through which reputation is built in wider socio-economic domains. High-tech designers are most likely aware that in getting WiFi makers to review and tinker with their product it will potentially reach new markets. The many IT business logos on the group's website and posters produced for public events, present further examples of similar collaborative relationships. Air-Stream promotes local businesses to its members, negotiates special discounts with retailers and welcomes business people as members. These relationships illustrate the indivisibility of the membrane that divides the expert from the amateur. Haring describes a similar overlap between amateur hams and professional radio operators:

> Hobbyists publicly promoted ties to the electronics industry to enhance their reputation for technical mastery. On the job, hams invoked the amateur persona. The particular styles of technical knowledge and practice associated with amateurs, hobbyists claimed, carried over into paid occupations. By this logic, professional success stemmed from amateur status, completely contradicting the usual meaning of amateur. (2007:89)

Like the trees, bugs and weather, I detailed in Chapter 5, WiFi makers find ways to work with and enfold local ISPs into their network. In the same way that traditional suburban boundaries of buildings and streets are largely irrelevant to wireless networks, the conventional divisions between professional and amateur appear not to apply. Just as Wyatt reminds us that 'digital exclusion does not always mean social exclusion'

DOI: 10.1057/9781137312532.0013

(2008:11), Air-Stream provides an example of homebrew WiFi that is not always situated in opposition to professional commercial practices. Their everyday practices permit other ways of engaging with conventional large-scale actors in existing technological landscapes.

A related example is provided in English-Lueck's (2003) study of New Zealand's high-tech industry. Known more for studying the cultures of Silicon Valley in California, she turned her attentions to New Zealand to explore its role as a silicon producer in the global economy where she examined the legitimacy of a local hands-on approach in high technology production.

> The last stop out before Antarctica, New Zealand has created a narrative around being at the 'ends of the Earth'. A tolerance for quirkiness is something that informants viewed as integral to New Zealander's ability to innovate. Niche research and development are key to New Zealand's place in the global silicon network. (4)

English-Lueck argues that New Zealand's approach emerges from an aptitude for local ingenuity, adaptive reuse and problem solving, all highly regarded attributes in the global technology marketplace. New Zealand's 'culture of innovation' is necessitated *and* enabled by isolation, distance and space and materials at hand. She argues that innovations do not have to be completely revolutionary or new to be valuable, instead value is perceived in unique re-combinations of existing materials and problems. Crucially, what this work demonstrates is that homebrew approaches to high-tech are highly regarded in global marketplace.

'Dodgy geezers' or innovative problem solvers?

Even though WiFi makers do not need to be professional to look professional, developing and maintaining reputation is critical to the group. Craig is often the first one to respond to problems on the network because he believes the ramifications of not looking professional would damage the group's standing in the community and in turn damage the network.

CRAIG: If people are connected to it and want to use it and they are complaining, of course I want to fix it straight away because I want them to tell their friends that Air-Stream is good. I don't want anyone bad mouthing things when we are trying to grow it, because we get enough of that from the commercials [ISPs]. Certain cowboys badmouth Air-Stream all the time.

DOI: 10.1057/9781137312532.0013

Craig's comments reveal not all relationships with local ISPs are positive. While people from business were encouraged to attend WiFi meetings and become members, some were less open about their agendas.

TIM: It was a bit snakey how they went about it. They actually came and talked to us before they went ahead and released it [new WiFi business] publicly. So they kind of picked our brains about how they should go about this but it didn't work. It flopped. Because the market they were targeting weren't interested in what they were offering.

As discussed in Chapter 6, some ISPs view WiFi makers as 'polluting' the shared spectrum. In turn, WiFi makers view and discuss at length some ISP's work as irresponsible, amateur and technically inept. In particular, they regularly critiqued the one-size-fits-all approach of installing inappropriately powerful technologies in places where they were not needed:

'Dodgy geezers.'
'They are cowboys.'
'And that thing [antenna] will burn birds. It's incredible.'
'It's like using a sledgehammer to whack a walnut.'
'It's like using monster trucks for when you need bicycles.'

These examples highlight how WiFi makers value their ability to deftly customise antennas to accommodate the subtle nuances of the materials at hand, the location, weather conditions, council restrictions, fauna and anticipated use of the technology. Calling ISPs 'cowboys' and 'dodgy' represents them as amateur and unprofessional, turning the tables on conventional assumptions of homebrew high-tech.

English-Lueck's (2003) juxtaposition of 'quirkiness' in the field of the 'big' science of silicon production dispels the idea that homebrew is incommensurate to high-tech. The importance of her work lies not only in recognising technology makers who innovate outside of large-scale institutions but in acknowledging unconventional methods and practices. In particular she describes how 'a rhetoric of frontier inventiveness is imbedded in the New Zealand idea that anything can be fixed with #8 fencing wire' (4). This example is particularly relevant both as a tool and as a metaphor for resourcefulness because WiFi makers have their own version in sticky tape. Sticky tape represents an important way of re-imagining how innovation and inventiveness might happen in the suburbs of Australia.

DOI: 10.1057/9781137312532.0013

Towards a sticky tape technology culture

> A cordless drill, screwdrivers, hammer, spanners, a silicon gun, cable ties, sticky tape and pliers lay strewn in the shed, on the grass outside, across the ping-pong table and wooden backyard garden furniture. I ask Simon and Dave which tools they most use: 'Sticky tape!'
>
> Two out of three antennas that Ben shows me are sticky taped in some way to keep them from falling apart or slicing his fingers.
>
> After showing photos and telling the monthly meeting about the installation of a new antenna on a challenging site, Paul says, 'It shows you don't need flash stuff, you just need sticky tape'.

It was easy to overlook the ubiquitous presence of sticky tape in the first few months of fieldwork. Other seemingly more exotic ethnographic objects and practices initially distracted me. Yet, sticky tape was stubbornly present; in backyards, sheds, pockets, kitchens, toolboxes, cars and backpacks. It was a key actor in interactions, demonstrations and experiences of how Air-Stream members made WiFi. Critically though, this mundane and ordinary artefact was not only an accessible and cheap material highly valued by members in their everyday situated practice but also when it was not actually used it was evoked as a way of working.

The theoretical and methodological ubiquity of sticky tape reflects how innovations do not have to be revolutionary or new. Value is perceived in re-combinations, re-interpretations, of existing materials and problems. Sticky tape also attracts a range of people who would normally not adhere to technology. In this way it complements the DiT approach, epitomising an experimental hands-on homebrew high-tech culture; an ability to bring together and make sense of heterogeneous networks of human and non-human actors. Sticky tape is an apt metaphor for this culture of WiFi makers because it evokes a particular method of binding. A key theme that has emerged throughout this book is WiFi makers' comfort with instability. It manifests in multiple interpretations of connectivity, technicalities, membership and even the activities of local birdlife and the weather. There is no single or dominant version of the network, style of stories or the group's visual culture. Instead, members collectively attach themselves to contingent assemblies of multi-dimensional representations. As a result, WiFi as it is made by backyard technologists fits with the idea of sticky tape that temporarily holds things in place only to be removed and re-stuck again in a different configuration.

DOI: 10.1057/9781137312532.0013

The concept of a sticky tape technology culture recalls what has been argued in STS about binding agents. In engineering, Henderson (1999) describes visual communication as 'glue'. In science, Latour's (1990) concept of 'circulating reference' highlights the cascading 'chains' of representations, comprehensively layered in cascades, each one linked to the next in line. These representations bind people, objects and knowledge together. They are held together by the stickiness of their visual culture. But these particular adhesives – glue and chains – are characteristically not temporary or mutable. These binding agents are evocative of a more permanent style of bond between objects and the people who make and use them. The ubiquity of sticky tape revealed the way members produce technologies that everyone can get their hands into; it helps them evade many of the restrictions presented by black box technologies.

Air-Stream's cultural practices and representations also stick people and objects together, but as per the character of sticky tape, the bond is differently comprised – it is a temporary fixture. People are not required to subscribe to the same structures or ways of working as anyone else. Plans are not firm. Ideas are not unyielding. Just like sticky tape, the group does not so much impose itself upon members or the technology as enable the constant development of a type of contingent practice. The ways in which sticky tape is used to hold together pieces of WiFi also makes it an important component of modding and making-do and is emblematic of a certain kind of Australian homebrew high-tech agility.

This chapter signals how backyard tinkering can result in the production of a sophisticated wireless technological product, previously an exclusive enterprise of large commercial or state governed infrastructural bodies. It bridges the distinct spheres of DiY and professional, with homebrew high-tech. As English-Lueck (2003) demonstrated in her New Zealand innovation studies, the essence of innovation is not necessarily always about something new. It can also be about recognising the value of mundane, overlooked things. Homebrew high-tech brings into being a socio-technology that is unique, locally constituted and imbued with the subjectivity of the maker/s. Making-do and modding incorporates much more than simply getting by or surviving. It is also about ingenuity and innovation. Drawing on the ubiquitous presence of sticky tape, both as a everyday tool and as an evocation of a way of working, signals a way of making sense of co-existing, overlapping, contingent findings. It can be seen as emblematic of WiFi makers' openness to ideas, collaborative ways of working, acceptance of mistakes as part of the process and

DOI: 10.1057/9781137312532.0013

ability to respond and adapt to constantly changing conditions. This approach is not a consequence of a fragile technology, the elastic nature of the group or an unpredictable environment but rather is deliberately produced, critical to how members innovate and expand the network.

DOI: 10.1057/9781137312532.0013

9

Do-it-Together Technology Cultures and Other Conclusions

Abstract: *As per the nature of the collective, multi-dimensional and at times messy practice of making WiFi, there is no single neat or narrow conclusion. This final chapter begins by returning to the field to contemplate the changes in Air-Stream practice and discussing the reasons for and results of socio-technical change. I draw attention to key themes emerging throughout the book relating to connectivity, visual culture and Do-it-Yourself (DiY) practice. Then, looking to the future, I conclude by speculating on the potential wider application of a Do-it-Together (DiT) approach in materializing other seemingly complex and complicated innovation processes and systems, and asking what other extraordinary things can ordinary people make in their own backyards.*

Jungnickel, Katrina. *DiY WiFi: Re-imagining Connectivity*. Basingstoke: Palgrave Macmillan, 2014. DOI: 10.1057/9781137312532.0014.

From Do-it-Yourself to Do-it-Together and beyond

I returned to Adelaide in 2013 to meet with WiFi makers, check references and technical specifications and discovered a different culture. Several members had left and were devoting their spare time, money and backyards to new technical hobbies such as Dorkbot,[1] Hackerspace[2] and the newly opened Adelaide Fab Lab. Movement in and out of the group, as I have illustrated throughout the book, was not in itself unusual. However, this exodus marked a distinct cultural shift born of technological change. Towards the end of my fieldwork in 2008, WiFi makers were beginning to note an increasing availability of low-cost, high-quality off-the-shelf equipment, which was lessening the need to mod and make-do. The following years witnessed a significant change in the socio-technical makeup of the group:

RON: They're not so much focused on the DiY anymore. They're more about how much data I can transfer from my house to my friend's house. Because the internet is still bi-directional, you have a fast download and a slow upload. And that's what a wireless network overcomes. And the equipment is cheap and all the same sites that allow it to go from one side of Adelaide to the other side of Adelaide are still there from all that work. They're upgrading them. So it's a different culture. In the past I would work very hard to document for public information sharing 'This is how you do it' and 'This is where you do it'. There are reasons that have changed that. One is the cheapness and availability of equipment, so nobody needs to know how you enclosed your circuit board in a waterproof case anymore or built a power over ethernet system. You don't need to know that stuff anymore.

All community groups mature and change. Haring closely documented a technical shift in ham radio culture which saw a similar increase in off-the-shelf devices, which 'unwittingly decreased satisfaction and learning by doing' (2007:148). In response, many ham operators returned to older, vintage equipment to remain connected to their technical culture and community values. This is not the case with Air-Stream. Although access to the internet is faster, as Ron notes, it remains bi-directional; it is still easier to download than upload which continues to privilege consumption over contribution. Remaining WiFi members and new recruits continue to provide alternatives to these restrictions, by way of a customised, independent and deeply local wireless network. Embracing technical change enables them to further hone the network, making it

DOI: 10.1057/9781137312532.0014

run better and faster. A by-product of the shift is fewer dramatic stories to tell and pictures to share, with unique mods becoming less frequent as antennas standardise. Nevertheless, network coverage and membership continues to grow in size and strength, providing yet another example of how makers enfold a potential 'threat' into their everyday practice and carry on making WiFi. Similarly, the range of new hobbies absorbing ex-WiFi makers signals a continued interest in collective technology practices involving a range of materials, open source practices and collaborative contexts. All, in their own way, are continuing to Do-it-Together (DiT).

As discussed in Chapter 2 and exemplified here, the digital techno-logical landscape is never static. The only constant is change. Since 2001, speed, access and choice of commercial broadband internet provision and services in Australia have significantly increased. Yet the nature of connection has not. Consumers connect to providers, then to the inter-net and finally with each other. The disparity between up and download speeds persists, constraining creative engagement not only *on* the inter-net but *with* the very nature of its infrastructure. As long as the internet remains mediated by unyielding hierarchical commercial models that situate and isolate subscribers at the bottom of a top down pay-plug-and-play consumer rather than a more flattened collective and collaborative contributor model, then stories like Air-Stream's will remain timely and timeless.

Air-Stream's aim has never been to simply provide access to *the* internet. If it were they would have dispersed years ago. They remain bonded by the desire to make connections and connections are infinite. This is why the group continues to exist and expand despite the relentlessly shifting technological context. Stories in the book provide ample evidence that it is the group's ability to deal with constant indeterminacy and multiple realities that affords it durability. They ride the waves of technical change, push at the limits of a technology, dream up new ideas and attempt to realise them. In doing so they bring to life the promise of a deep engage-ment with a collaborative contribution based socio-technical model. In the move from bedroom to backyard, individual to collective and community to global, Air-Stream enacts the shift from distributed lone makers to a collective whole. At its core, DiT culture taps into the impulse to harness shared skills, resources and ideas for seemingly infinite appli-cation. As Thomson (2007) writes, 'The answer's in our own backyards'.

DOI: 10.1057/9781137312532.0014

Closely examining how a group of people demystify a sophisticated and largely invisible technology in their homes and spare time by rendering it visible, using mundane materials and openly sharing knowledge gives rise to a range of questions that go beyond the immediate subject of the book. The potential wider application of a DiT approach in materializing seemingly complex and complicated technological innovation processes and systems triggers the questions: If broadband wireless internet can be made by without specialised equipment or training in backyards what else can be learnt from making things visible? What modes of support might catalyse larger scale community DiT technology cultures? Who is currently excluded and could be involved? What other kinds of high-tech might be homebrewed?

At the time of writing, Adelaide had launched Australia's first Fab Lab, a small scale, open source fabrication laboratory bearing the tagline: 'Think it, Make It, Share It'. Only one of two in the southern hemisphere, Adelaide's Fab Lab continues a concept started at Massachusetts Institute of Technology's Media Lab for the purpose of providing community members with space and access to sophisticated equipment with which to invent, make and share new things. I have not investigated this new socio-technical community in detail, however even a cursory analysis reveals remarkable similarities between the Fab Lab Charter and Air-Stream core values.[3] The expansion of collective technically oriented making spaces like Fab Lab and the surge of public excitement that surrounds them serves to strengthen the significance of the Air-Stream story. The fact that Air-Stream has been running for over a decade signals the pioneering role backyard technologists have played in creating a culture where spaces like Fab Lab are possible. Air-Stream provides a working example of commercial providers and individual makers operating in tandem in mutually beneficial ways. They demonstrate how makers, if given access to black box technologies, will tinker with their contents and imagine as well as generate applications that extend beyond what the original designers had in mind. They also fortify the value of open source practices, illustrating how sharing knowledge and tools, revealing mistakes as well as achievements and pooling skills, time and resources fosters a culture that works *with* mess, interruption and diversity, which in turn makes a group/idea resilient and responsive to change. Air-Stream is a case study of ordinary people making extraordinary things in everyday places.

DOI: 10.1057/9781137312532.0014

An Australian WiFi made by backyard technologists

In answer to the question what would different versions of WiFi look like, this book has described an Australian WiFi network, or to be more succinct an Adelaide one between 2006 and 2009. Just as the internet is not a single universal entity but one of many versions that co-exist, this study provides evidence that WiFi is similarly multiple. Many actors particular to Adelaide are integral to the making of WiFi. Flyscreen, sticky tape, biscuit tin boxes and Hills Hoist clothes line wire are just a few suburban actors that found their way into a developing digital technology. I described the barbeque as a quintessential social and collaborative framework that gave shape to WiFi events. Likewise, birds, local winds, long summer months, eucalyptus trees and even bugs were incorporated in daily process of making the network. 'Ournet not the internet' is a local made-to-measure version of the internet, indelibly shaped by sensitivity to the local landscape and its inhabitants. Moreover, drawing on Miller and Slater (2000) who argued that Trinidadians think of the internet as Trinidadian, this book revealed an Australian propensity for WiFi. This is because tinkering and backyard installations fit with local culture in the way the internet with its support for chatting and identity politic fitted in Trinidad. The implications of an Australian WiFi suggest the presence of alternate versions of WiFi in other countries, raising the questions: What would British WiFi look like? How might French WiFi compare to Portuguese WiFi?

However, it is not simply the presence of local actors that constitute a local WiFi, but *how* they come together that is important. This marks the study's larger contribution to the study of wireless digital technologies. A central theme throughout the book is how makers imbue a Do-it-Yourself ethic but do not do it alone, they Do-it-Together. Throughout are examples of DiT practice – the sharing of skills, materials, sites, time, money and ideas – producing a technology borne of give and take. The community WiFi network is dialogic, continually informing and informed by interchanging actors. Contribution is axiomatic. In return WiFi makers learn how to deal with challenging problems, develop knowledge through mistakes, cement social ties and in the process, open up new landscapes of connective possibility. Stories throughout this book illustrate how WiFi does not slot easily into suburbia; it often breaks and needs adapting to an ever-shifting ecology. Yet makers rarely became anxious about things that did not fit, occasionally

DOI: 10.1057/9781137312532.0014

disappeared or failed to work. This is because they are empowered to fix it themselves. Rather than wait for a specialist technician, they mod and make-do with a seemingly indefatigable array of materials at-hand, which delivers a sense of power because they are certain of their ability to tackle problems. Mods and modding are part of a backyard technologist's toolkit. Mods are a unique approach to making-do, a local style of adaptability and ingenuity borne of a unique constellation of taxing conditions, colonial heritage, a range of local actors and inappropriate tools and equipment. Modding involves putting things together in new ways, assembling human and non-human actors into newly configured heterogeneous networks. Vital to experimentation, this practice emerges from a deep understanding of the properties of material, place and personal skills. Mistakes and things that break are not failures but rather serve to affirm members' curiosity and ingenuity. This flexible and responsive approach means makers adapt to a wide range of instabilities, customising responses as well as technologies and dispelling the on-size-fits-all leitmotif that characterises commercial systems. In particular, examples provided in Chapters 7 and 8 illustrated how mapping contemporary technological imaginaries onto historical ones helps to generate original ways of understanding current infrastructural issues and pave the way for dealing with future ones.

The book has also described the dual role of backyards and sticky tape as emblematic of a homebrew high-tech approach. The study took these mundane entities seriously. It moved them from in and behind the house out into a global context by positioning them as vital sites/tools for wireless technological innovation symbolic of a hands-on, resourceful and collective approach. Backyard technologists offer a distinctive way of socially imagining a new technology; materially thinking through ideas with fresh eyes on at-hand materials. Like backyards, sticky tape emerged in this book as an everyday tool and evocation of a way of making sense of co-existing, overlapping, contingent findings and of attaching a diverse range of people together. Critically, sticky tape is not irrevocably binding but rather temporarily holds stuff in place, enabling things to be removed and re-stuck again in alternate configurations. It enables makers to re-imagine how things might be.

While this book illustrates how extraordinary technologies *are* within the reach of ordinary people, it also questions the nature of the 'ordinary', asking who can and cannot participate in this grassroots technology culture. As open as the group was to new members, it remained

DOI: 10.1057/9781137312532.0014

predominantly white, male and middle-class. I did not set out to explore the role of women in WiFi, nor directly ask 'where are the women?' Instead, given their overt absence, gender became a subtext through which other actors and their actions were rendered all the more present. What these stories show is that WiFi is not just what Australians do, but particularly what Australian men do. The absence of women in the group served to normalise the idea that WiFi seemed to fit more comfortably or 'naturally' with men. This is despite the fact the group represents itself as a 'community' group and exhibits fewer of the traditional barriers to entry that inhibit access by women. Following Faulkner's (2000) example in engineering, I explored what made WiFi 'stick' to men. I also described how women were involved in the network, essential to its success, yet often hidden behind the scenes. This is not to say that only men make new wireless digital technology in Australia. But for Air-Stream, women played a supporting role. In some ways, what the study of Air-Stream reveals is while new kinds of technological imaginings are possible, some require even more resourcefulness and ingenuity to bring to life.

How WiFi 'might have been otherwise'

> Technologies do not, we suggest, evolve under the impetus of some necessary inner technological or scientific logic. They are not possessed of an inherent momentum. If they evolve or change, it is because they have been pressed into that shape. But the question then becomes: why did they actually take the form that they did? (Bijker and Law 1992:2)

A major motivation of the book has been to offer fresh perspective on debates about the role of local culture and grassroots practice in the shaping of technological infrastructure. I set out to look at a new digital technology from the ground up, to ask why it took the shape it did and offer evidence of how it 'might have been otherwise' (Bijker and Law 1992:3). Despite widespread interest and adoption, scholarly attention has lagged behind media rhetoric and commercial representations of WiFi.

Stories in the book attempt to fill this gap by drawing attention to the importance of mundane and ordinary practices and architectures of suburban life in the development and understanding of new wireless digital technologies. They illustrate how WiFi makers re-inscribe wireless broadband technology with new meanings and re-imagined possibilities

DOI: 10.1057/9781137312532.0014

of use. For instance, although Ron's initial interest was catalysed by a practical desire to share an internet connection between two of his businesses, he soon became passionate about WiFi's socio-technical possibilities:

> I often get really excited in that there are so many things you can do with this technology. I've gone to see councils and I've gone to see people in state government. And I say, 'Hey you can do this' and 'We are doing all this' and 'It's so great' and 'You can have kiosks around the community' and 'People can access information' and 'You don't have to pay internet costs' and 'You can link your council billing systems so it's there 24 hours a day' and 'People can go and pay their bill through your little teller that is out there in the street or in the local mall'. Or 'You can have local telephony for people who are homeless' and things like that. 'You can do all these things at very low cost'... 'You could have access points with information for the travellers with information about hotels and all the local business'. Surely in local towns you'd want to encourage the local economy? But they were very much interested in how do we get these communities not just connected but how do we get them connected through ISPs.

Ron articulates some of the possibilities he envisions for circumventing traditional one-way relationships borne of a transaction for services. He imagines a constellation of new ways local councils could connect and interact with constituents using WiFi, including giving voice to those traditionally silenced by new technologies, the homeless. However, these organisations dismiss the idea, resorting instead to ISPs as the mediator. In answer to Green and Harvey's (1999:12) question 'what *disconnections* are entailed in connecting', here the council's narrow version of connectivity distances them from a direct relationship with constituents. They remain wedded to conventional consumer models, ignoring the larger question of what people are connecting to and why and what other connections might be possible. In this system, commercial mediators, in the form of ISPs who are in turn shaped by larger regulatory systems, prescribe what connectivity is, shaping and limiting how and in what ways we communicate, create meaning and relationships.

KAT: What is the Air-Stream network to you?

SIMON: A group of people who have come together to create something from an idea and who are continuing this idea and incorporate more and more people. Behind all of this I see the people more than the nodes and the access points. Ben up at Pasadena. That's not Air-Stream up at Pasadena, that's Ben's setup there. People connect to Ben. They don't connect to Pasadena. Yes, they have

DOI: 10.1057/9781137312532.0014

the equipment to connect but it's them that we are talking to and interacting with ... we've all got our own worlds at our places and with Air-Stream you can connect them.

Illustrated throughout the book and especially here in Simon's comment are examples of how the group attaches to each other first and then to the network. One of the aims of the research was to examine and intervene in conventional understandings of technological connectivity. In media and academic scholarship it has been largely explored as a matter of access, pressure or choice, which have shaped understandings of connectivity. Debate has also been framed in relation to government and powerful telecommunications organisations who dominate technological landscapes and discourse in Australia. This study reveals the possibility of multiple, shifting and uncertain connections as a result of the 'maybe', 'kind of', 'yesterday, but not today' and 'only if it does not rain' links that frequent this book. These accounts challenge the idea that technological connectivity is in any way definite or certain; being disconnected, unconnected or never being connected is the norm. Instead, makers are brought together for the purpose of *getting* connected, which supports a range of creative connective possibilities. A core feature of Air-Stream's connectivity is its local-ness. It connects people as close as a few blocks of one another and as far as a few suburbs apart. As a result, it presents a striking contrast to how the internet has been positioned as a tool for globalisation.

The fact that connectivity and time are not contiguous intervenes in the idea of WiFi as 'anytime' and 'anywhere' put forth in the rhetoric that surrounds it. The idea of WiFi's 'always on' connectivity is all about temporality. It relies upon the intersection of time and space; one connects to something and one expects a response. In disaggregating the certainty of connection (as well as the internet) from WiFi, this study presents an alternative to the idea of uniform systems and practices that dominate commercial provision. Air-Stream's network is not bought and plugged in but rather is uniquely made and customised to each location, shaped by available materials and skills. Each point in the network is distinct as illustrated by Reg's tree antenna in Chapter 5 and Dave's shed antenna in Chapter 7. Further still, there is a *lot* of work involved in making WiFi, much of which appears disproportionate to the return. Stumbling, for instance, in Chapter 6 reveals how makers think little of spending hours on suburban rooftops in the scorching sun simply 'looking around'. These examples would be considered inefficient at best and failures at worst if read in relation to the desire for, and expectation of, constant

DOI: 10.1057/9781137312532.0014

connectivity touted by conventional technological models. Instead, they reflect what Thomson has written about Australian shed and tinkering culture:

> This aptitude for nifty solutions with a length of fencing wire, a hammer and a piece of 4" × 2" timber is strongly imagined and widely felt to be some sort of competitive advantage. People take pride in such skills even when more sensible solutions may be available. (2002b:8)

This book is also about what we can and cannot see, how we make things known and who is present and absent in these representations, stories and imaginaries. Inspired by Star (1999), this study set out to explore a technological infrastructure commonly overlooked and undervalued. Connection is not static and representations, in multi-dimensional form, are central to how Air-Stream make WiFi. Yet, equivalences are conceived not by transforming raw materials into infinitely comparable and combinable representations, but by gathering together many different multi-dimensional forms that remain largely unfixed or temporarily sticky taped together. In this way, Air-Stream's representational culture echoes the co-located, overlapping and occasionally contradictory characteristics of its technical network. It sticks people together, but not in chains or glue as identified in STS. Air-Stream's tactics, characterised by mods and making-do, all point to more temporal adhesives that adapt to changing conditions.

Visual culture is defined as the way 'a culture *sees* the *world* and makes it visible' (Latour 1990:30; emphasis in original). Although this suggests an infinite and dynamic elucidation of ideas, conventional visual typology of knowledge objects has largely been confined to text, maps, sketches, drawings and images: a two-dimensional vocabulary for a multi-dimensional world. WiFi makers' spectrum of expression takes a myriad of forms: drawings, maps, diagrams, photos, demos, bodies, nodes, stumbling 'rigs', websites and blogs, IM, modems, tools and branded merchandise. To make sense of this complex array of objects required an ability to analyse across three dimensions, two dimensions, single dimension and, on occasion, in the form of radio signals, no dimension. Rather than imposing upon or demanding conformity of its members, Air-Stream's visual culture gathers together and forges connections between incommensurate actors such that they make sense not to a select few but to a diverse and distributed audience located inside *and* outside the group. Although public exposure presents threats to the

DOI: 10.1057/9781137312532.0014

network, such as the thefts that opened this book, WiFi makers accommodate and build them into their network, thus diffusing their ability to destabilise the system. The technical shifts discussed earlier and the way they enfolded me and my work into their visual culture is evidence of this systematic practice. The way the group's DiT imbued my methodological practices further signals the impulse to attach to these kinds of cultures and the autonomy for application and interpretation.

By way of a final point, what this book has signalled is the critical importance of paying attention to dynamic examples of small technological cultures such as community WiFi groups that are all too often easy to trivialise, ignore or overlook. Rendering them visible, locating them in rich cultural contexts and pointing to their role in 'engineering alternatives' help us to imagine other ways of being, connecting and knowing. The use of the word 'make/r' as a descriptor of people and practice throughout the book is, for this reason, deliberate. Air-Stream's WiFi network is suspended in the state of making. As a result, it is never finished. It can always be something else, something better, faster, more reliable and different to *the* internet. Makers are not content with how it is, but constantly seek what it could be. As the Fab Lab example illustrates, once a culture of making is established in a place, it quickly grows in new directions with individuals imagining other things they might tinker and mod. Perhaps more broadly what the Air-Stream story offers is a way to think not only about how WiFi 'might have been otherwise' but how we might apply these imaginings more extensively to the world around us.

Ron: There's an oil well going in, in South Australia. [Who says] you can't go up there and chain yourself to the fence. You *can* do that. Go and do it, 'cause I'm damn going to. People need to do stuff and just letting it all wash over you and consuming...I think that's what the whole DiY culture is about...empowerment. Taking control of your life and doing stuff and being part of the community.

Notes

1   See organizations affiliated to the website http://dorkbot.org around the world that cater to 'people doing strange things with electricity'.
2   See http://hackerspaces.org community operated places where people meet, share tools, ideas and resources and work on projects.
3   See http://fablabadelaide.org.au/what-is-a-fab-lab/fab-lab-charter/.

DOI: 10.1057/9781137312532.0014

Bibliography

ABS (2012) *Internet Activity, June Quarter*, Report No. 8153.0, 9 Oct, Accessed 04.12.12, Available at http://www.abs.gov.au.

ABS (2011) *Australian Regional Population Growth*, Report No. 3218.0, 31 June, Accessed 10.10.11, Available at http://www.abs.gov.au.

ABS (2008) *Australian Social Trends*, Report No. 4102.0, July, Accessed 12.08.08, Available at http://www.abs.gov.au.

ABS (2001) *Internet Activity, June Quarter*, Report No. 8153.0.0, 27 Sept, Accessed 10.04.06, Available at http://www.abs.gov.ac

Allen, M. and J. Long. (2004) 'Domesticating the Internet: Content Regulation, Virtual Nation-Building and the Family'. In G. Goggin (ed.), *Virtual Nation: The Internet in Australia*. Sydney: University of New South Wales Press, pp. 229–241.

ALP (2007) New *Directions for Communications: A Broadband Future for Australia – Building a National Broadband Network*, ALP, Canberra. Accessed 10.10.08, Available at http://www.iia.net.au.

Back, L., Halliday, P., Knowles, C., Wakeford, N. and A. Coffey. (2008) *Live Sociology: Practising Social Research with New Media*, Research Workshops, Goldsmiths College, University of London, Accessed: 10.07.08, Available at http:www.goldsmiths.ac.uk.

Battersby, L. (2013) 'Speed Check: What NBN Speed Promises Really Mean', 7 Aug, *Sydney Morning Herald, IT Pro*, Accessed 10.9.13, Available at www.smh.com.au.

DOI: 10.1057/9781137312532.0015

BBC News (2005a) 'Wireless Broadband Goes to Church', 31 May, Accessed 27.11.05, Available at http://news.bbc.co.uk.

BBC News (2005b) 'Rail Stations to be WiFi Enabled', 10 May, Accessed 27.11.05, Available at http://news.bbc.co.uk.

BBC News (2004) 'Surf the New While Surfing the Waves', 18 June, Accessed 27.11.05, Available at http://news.bbc.co.uk.

Bell, G. and P. Dourish (2006) 'Back to the Shed: Gendered Visions of Technology and Domesticity'. *Personal & Ubiquitous Computing: At Home with IT*, 11:5, 373–381.

Bijker, W. E. (1995) *Of Bicycles, Bakelites and Bulbs: Toward a Theory of Social Change*. Cambridge, MA: MIT Press.

Bijker, W. E. and J. Law (1992) *Shaping Technology/Building Society: Studies in Sociotechnical Change*. Cambridge, MA: MIT Press.

Birmingham, J., Jack, R. I. and D. N. Jeans (1979) *Australian Pioneer Technology: Sites and Relics: Towards an Industrial Archaeology of Australia*. Melbourne: Heinemann.

Bollen, J., Kiernander, K. and B. Parr (2008) *Men at Play; Masculinities in Australian Theatres since the 1950s*. Amsterdam and New York: Rodopi.

Braue, D. (2010) *We Need the NBN Because Business Does*, 3 Aug, Accessed 12.3.12, Available at http://www.zdnet.com/we-need-the-nbn-because-business-does-1339304950/.

Bruns, A. and J. Jacobs (eds) (2006) *Uses of Blogs*. New York: Peter Lang Publishing.

Budde, P. (2012) *Australia – Telecoms Industry – Statistics and Forecasts*, Accessed 10.12.12, Available at http://buddle.com.au.

Burrell, J. (2009) 'The Field Site as a Network: A Strategy for Locating Ethnographic Research'. *Field Methods*, 21:2, 181–199.

Callon, M. (1986) 'The Sociology of an Actor-Theory-Network: The Case of the Electric Vehicle'. In M. Callon, L. Law and A. Rip (eds), *Mapping the Dynamics of Science and Technology*. London: Macmillan Press.

Cartwright, L. (1995) *Screening the Body; Tracing Medicine's Visual Culture*. Minneapolis: University of Minnesota Press.

Cockburn, C. (1983) *Brothers: Male Dominance and Technological Change*. London: Pluto Press.

Cockburn, C. and M. Ormrod (1994) *Gender and Technology in the Making*, London: Sage.

Dunbar-Hester, C. (2008) 'Geeks, Meta-Geeks, and Gender Trouble: Activism, Identity, and Low-Power FM Radio'. *Social Studies of Science*, 38:2, 201–232.

DOI: 10.1057/9781137312532.0015

Dunne, A. (1999) *Hertzian Tales: Electronic Products, Aesthetic Experience and Critical Design*. London: RCA CRD Research Publications.

Elder, C. (2007) *Being Australian: Narratives of National Identity*. Sydney: Allen & Unwin.

Ellis, R. and C. Waterton (2005) 'Caught between the Cartographic and the Ethnographic: The Whereabouts of Amateurs, Professionals and Nature in Knowing Biodiversity'. *Environmental Planning D, Society & Space*, 23:5, 673–693.

Emerson, R. M., Fretz, R. I. and L. L. Shaw (2011) *Writing Ethnographic Fieldnotes*, Second edition. London and Chicago: Chicago University Press.

English-Lueck, J. A. (2003) 'Number Eight Fencing Wire: New Zealand, Cultural Innovation and the Global Silicon Network', Adaptation of poster presented at the 2003 *Annual Meeting of the American Anthropological Association*, 20 Nov, Chicago, IL, Accessed 15.04.07, Available at http://www.sjsu.edu/depts/anthropology/.

Faulkner, W. (2007) '"Nuts and Bolts and People': Gender-Troubled Engineering Identities'. *Social Studies of Science*, 37:3, 331–356.

Faulkner, W. (2006) *Gadget Girls and Boys with their Toys: How to Attract and Keep More Women in Engineering*, Pamphlet, Accessed 10.1007, Available at http://www.esrcsocietytoday.ac.uk.

Faulkner, W. (2000) 'The Power *and* the Pleasure? A Research Agenda for "Making Gender Stick" to Engineers'. *Science, Technology & Human Values*, 25:1, 87–119.

Fetterman, D. M. (2010) *Ethnography: Step by Step*. Third edition, Applied Social Research Methods Series, 17, London: Sage.

Fisk, J., Hodge, B. and G. Turner (1987) *Myths of Oz: Reading Australian Popular Culture*. Sydney: Allen & Unwin.

Forlano, L. (2008) 'Anytime? Anywhere? Reframing Debates around Municipal Wireless Networking'. *The Journal of Community Informatics, Special Issue: Wireless Networking for Communities, Citizens and The Public Interest*, 4:1, Available at http://ci-journal.net.

Foth, M. (ed.) (2009) *Handbook of Research on Urban Informatics: The Practice and the Promise of the Real Time City*. Hershey, PA: Information Science Reference, IGI Global.

Foth, M., Forlano, L., Satchell, C. and M. Gibbs (eds) (2012) *From Social Butterfly to Engaged Citizen: Urban informatics, Social Media, Ubiquitous Computing, and Mobile Technology to Support Citizen Engagement*. Cambridge, MA: MIT Press.

DOI: 10.1057/9781137312532.0015

Fowler, B. and F. Wilson (2004) 'Women Architects and Their Discontents'. *Sociology*, 30:1, 101–119.

Garforth, L. (2011) 'In/Visibilities of Research: Seeing and Knowing in STS'. *Science, Technology & Human Values*, 37:2, 264–285.

Given, J. (2008) 'Australia's Broadband: How Big a Problem Is It?' *Media International Australia*, 127, 6–10.

Glaser, B. G. and A. L. Strauss (1967) *The Discovery of Grounded Theory: Strategies for Qualitative Research*. Chicago: Aldine Publishing Company.

Goggin, G. (2007) 'An Australian Wireless Commons?' In G. Goggin and M. Gregg (eds), *Media International Australia* Special Issues on Wireless Cultures and Technologies, pp. 118–130.

Goggin, G. (ed.) (2004) *Virtual Nation: The Internet in Australia*. Sydney: University of New South Wales Press.

Goggin, G. and M. Gregg (eds) (2007) *Media International Australia* Special Issues on Wireless Cultures and Technologies.

Graham, S. and N. Thrift (2007) 'Out of Order: Understanding Repair and Maintenance'. *Theory, Culture & Society*, 24:3, 1–25.

Green, S. (2000) Opening Plenary, *Virtual Society? Get Real! Conference*, 4–5 May, London: ESRC Programme: Virtual Society? The Social Science of Electronic Technologies.

Green, S., Harvey, P. and H. Knox (2005) 'Scales of Place and Networks: An Ethnography of the Imperative to Connect through Information and Communications Technologies'. *Current Anthropology*, 46:5, 805–826.

Green, S. and P. Harvey (1999) 'Scaling Place and Networks: An Ethnography of ICT "Innovation in Manchester"', Paper presented at *Internet and Ethnography Conference*, 13–14 Dec, Hull University, Available at http://les.man.ac.uk.

Gregg, M. (2011) *Work's Intimacy*. Cambridge, UK: Polity Press.

Gregg, M. (2007) 'Freedom to Work: The Impact of Wireless on Labour Politics'. In G. Goggin and M. Gregg (eds), *Wireless Cultures and Technologies*. Media International Australia: The University of Queensland, pp. 57–70.

Hall, T. (2007) *Where Have All the Gardens Gone? An Investigation into the Disappearance of Backyards in the Newer Australian Suburb*. Brisbane: Griffith University.

Hammersley, M. and P. Atkinson (2007) *Ethnography: Principles in Practice*. Third edition. Oxon: Routledge.

DOI: 10.1057/9781137312532.0015

Harding, S. (1991) *Whose Science? Whose Knowledge? Thinking from Women's Lives*. Open University Press.

Haring, K. (2007) *Ham Radio's Technical Culture*. Cambridge, MA and London, England: MIT Press.

Henderson, K. (1999) *On Line and On Paper: Visual Representations, Visual Culture, and Computer Graphics in Design Engineering*. Cambridge, MA: MIT Press.

Hess, D. (2002) 'Ethnography and the Development of Science and Technology Studies'. In P. Atkinson, A. J. Coffey, S. Delamont, J. Lofland and L. H. Lofland (eds), *HandBook of Ethnography*. London: Sage, pp. 234–244.

Hine, C. (2000) *Virtual Ethnography*. London and New Delhi: Sage Publications.

Hjorth, L. (2007) *Waiting for Immediacy Catalogue*. Seoul: Yonsei University. Copy provided by author.

Ito, M., Okabe, D. and M. Matsua (eds) (2005) *Personal, Portable, Pedestrian: Mobile Phones in Japanese Life*. Cambridge: MIT Press.

Jackson, S. (2006) 'Sacred Objects – Australian Design and National Celebrations'. *Journal of Design History*, 19:3, 249–255.

James, G. (2003) *Bridging the Global Digital Divide*. UK: Edward Elgar Publishing.

Johnson, C. (2008) 'Kevin Rudd and the Labor Tradition: Economy, Technology and Social Diversity', Accepted paper for *The Political Studies Association Conference*, 6–9 July, Brisbane, Australia: Hilton Hotel.

Jordan, T. (2002) *Activism! Direct Action, Hacktivism and the Future of Society*. Reaktion Books.

Jungnickel, K. (2013) 'Getting there . . . and back: How Ethnographic Commuting (by Bicycle) Shaped a Study of Australian Backyard Technologists'. *Qualitative Research* (published online, 4 Apr).

Jungnickel, K. (2010) 'Exhibiting Ethnographic Knowledge: Making Sociology about Makers of Technology'. *Street Signs; Centre for Urban and Community Research*, Spring. London: Goldsmiths.

Jungnickel, K. (2009) 'Makers', 'Mashers' & 'Mods': Grassroots technology practices', Research Exhibition at *The Sociological Imagination: A 50th Anniversary Celebration Conference*. London: Goldsmiths.

Jungnickel, K. (2006) 'Hacking the Home: Technological Tantrums and Wireless Workarounds in Domestic Culture'. *Wireless Cultures and Technologies Workshop*, 2 Dec. Sydney: University of Sydney.

DOI: 10.1057/9781137312532.0015

Jungnickel, K. and G. Bell. (2008) 'Home is where the hub is? Wireless infrastructures and the nature of domestic culture in Australia'. In M. Foth (ed.), *Handbook of Research on Urban Informatics: Community Integration & Implementation.* Hershey, PA: IGI Global.

Kleif, T. and W. Faulkner. (2003) '"I'm no Athlete [but] I can Make This Thing Dance!" Men's Pleasures in Technology'. *Science, Technology & Human Values*, 28:20, 296–325.

Koprowski, G. J. (2004) 'Wireless World: City Wi-Fi Networks Growing', *United Press International*, 11 Dec, Accessed 4.10.05, Available at http://www.upi.com.

Kotamraju, N. P. (1999) 'The Birth of Web Site Design Skills'. In P. Lyman and N. Wakeford (eds), *Analysing Virtual Societies: New Directions in Methodology. American Behavioural Scientist*, 43 (3): 377–391.

Latour, B. (2005) *Reassembling the Social: An introduction to Actor-Network-Theory*, Oxford: Oxford University Press.

Latour, B. (1996) *Aramis, or the Love of Technology*, Cambridge, MA: MIT Press.

Latour, B. (1992) 'Where are the Missing Masses? The Sociology of a Few Mundane Artifacts'. In W. E. Bijker and J. Law (eds), *Shaping Technology/Building Society: Studies in Sociotechnical Change.* Cambridge, MA: MIT Press.

Latour, B. (1990) 'Drawing Things Together'. In M. Lynch and S. Woolgar (eds), *Representation in Scientific Practice.* Cambridge, MA: MIT Press, pp. 19–68.

Latour, B. (1987) *Science in Action: How to Follow Scientists and Engineers through Society.* Massachusetts: Harvard University Press.

Latour, B. and Yaneva, A. (2008) 'Give me a Gun and I Will Make All Buildings Move: An ANT's View of Architecture'. In: R. Geiser (ed.), *Explorations in Architecture: Teaching, Design, Research.* Basel: Birkhäuser, pp. 80–89.

Latour, B. and S. Woolgar (1979) *Laboratory Life, the Social Construction of Scientific Facts.* London: Sage.

Law, J. (2004) *After Method: Mess in Social Science Research.* UK: Routledge.

Law, J. and V. Singleton (2005) 'Object Lessons'. *Organisation*, 12:3, 331–355.

Law, J. and A. Mol (2001) *Situating Technoscience: An Inquiry into Spatialities,* Centre for Science Studies, Lancaster University,

DOI: 10.1057/9781137312532.0015

Lancaster, Accessed 10.10.05, Available at http://www.comp.lancs. ac.uk.

Lessig, L. (2004) 'Free Culture: How Big Media Uses Technology and the Law to Lock Down Culture and Control Creativity'. *The Penguin Press*, Accessed 10.02.06, Available at http://www.free-culture.cc/.

Libbenga, J. (2003) 'Pocket Wi-Fi sniffers End Missing Hotspot Misery'. *The Register Online*, 18 Sept, Accessed 12.10.05, Available at http://wwtheregister.co.uk.

Lingel, J. and M. Naaman (2012) 'You should have been there, man: Live Music, DIY Content and Online Communities'. *New Media & Society*, 14:2, 332–349.

Lofland, L. H., Lofland, J., Snow, D. and L. Anderson (2004) *Analysing Social Settings: A Guide to Qualitative Observation and Analysis.* Fourth edition. Belmost, CA: Wadsworth Publishing Company.

Lury, C. and N. Wakeford (eds) (2012) *Inventive Methods: The Happening of the Social.* London: Routledge.

Lynch, M. and S. Woolgar (eds) (1990) *Representation in Scientific Practice.* Cambridge, MA: MIT Press.

Mackenzie, A. (2007) 'Wireless Networks and the Problems of Overconnectedness'. In G. Goggin and M. Gregg (eds), *Wireless Cultures and Technologies.* Media International Australia: The University of Queensland, pp. 94–105.

Mackenzie, A. (2005a) 'Untangling the Unwired: The Cultural Implications of Wireless (Wi-Fi) Infrastructures'. *Space and Culture*, 8:3, 269–285.

Mackenzie, A. (2005b) 'From Cafe to Parkbench: WiFi and Technological Overflows in the City'. In M. Sheller (ed.), *Technological Mobilities.* London and New York: Routledge.

MacKay, H. (2008) 'Suburbia, Understanding Australia – Program 5', Interviewed by Sue Slamen, Radio Australia, *Australian Broadcasting Corporation*, Accessed 10.09.08, Available at http://www. radioaustralia.net.au.

Marcus, G. (1998) 'Ethnography in/of the World System: The Emergence of Multi-sited Ethnography'. In *Ethnography through Thick and Thin.* Princeton, NJ: Princeton University Press, pp. 79–104.

Marks, K. (2004) 'Aussies Mourn Loss of Backyard to Urban Sprawl'. *The Independent*, 25 July, Accessed 06.06.08, Available at http://www. independent.co.uk.

DOI: 10.1057/9781137312532.0015

McIlwee, J. and J. G. Robinson (1992) *Women in Engineering: Gender, Power and Workplace Culture*. Albany: State University of New York Press.

McKay, G. (ed.) (1998) *DIY Culture: Party & Protest in Nineties Britain*. London and New York: Verso.

Meikle, G. (2004) 'Networks of Influence: Internet Activism in Australia and Beyond'. In G. Goggin (ed.), *Virtual Nation: The Internet in Australia*. Sydney: University of New South Wales Press, pp. 73–87.

Michael, M. (2006) *Technoscience and Everyday Life, The Complex Simplicities of the Mundane*. London: Open University Press.

Michael, M. (2000) *Reconnecting Culture, Technology and Nature: From Society to Heterogeneity*. London: Routledge.

Miller, D. and D. Slater (2000) *The Internet: An Ethnographic Approach*. London: Routledge.

Mol, A. (2002) *The Body Multiple: Ontology in Medical Practice*. Durham, NC and London: Duke University Press.

Morris, S. (2004) 'Make New Friends and Kill Them: Online Multiplayer Computer Game Culture'. In G. Goggin (ed.), *Virtual Nation: The Internet in Australia*. Sydney: University of New South Wales Press, pp. 133–144.

Mvers, N. (2008) 'Molecular Embodiments and the Body-work of Modeling in Protein Crystallography'. *Social Studies of Science*, 38:2, 163–199.

Orr, J. (1996) *Talking about Machines: An Ethnography of a Modern Job*. Ithaca and London: Cornell University Press.

Oudshoorn, N. (2012) 'How Places Matter: Telecare Technologies and the Changing Spatial Dimensions of Healthcare'. *Social Studies of Science*, 42:1, 121–142.

Pease, B. and K. Pringle (2001) *A Man's World; Changing Men's Practices in a Globalised World*. New York: Zed Books.

Personal Telco (2012) 'Wireless Communities', *Portland, Oregon Volunteer Community Wireless Group*, Accessed 2.01.13, Available at https://personaltelco.net/wiki/WirelessCommunities.

Powell, A. (2012) 'WiFi Publics: Defining community and technology at Montreal's Île Sans Fil'. In A. Clement, M. Gurstein and G. Longford (eds), *Connecting Canadians: Investigations in Community Informatics*. Edmonton, Canada: Athabasca University Press, pp. 202–217.

Rennie, E. and S. Young (2004) 'Park Life: The Commons and Communications Policy'. In G. Goggin (ed.), *Virtual Nation: The*

DOI: 10.1057/9781137312532.0015

Internet in Australia. Sydney: University of New South Wales Press, pp. 242–257.

Rudd, K. (2009) *Transcript of Prime Minister Kevin Rudd's Press Conference on the National Broadband Network.* Australian Parliament House: Canberra, 7 April, Available at http://www.theaustralian.com.au.

Sandvig, C. (2004) 'An Initial Assessment of Cooperative Action in Wi-Fi Networking'. *Telecommunications Policy,* 28:7/8, 579–602.

Schwartz-Cowan, R. (1983) *More Work for Mother: the Ironies of Household Technology from the Open Hearth to the Microwave.* New York: Basic Books.

Searle, D. (ed.) (1997) *Gathering Force: DIY Culture – Radical Action for Those Tired of Waiting.* London: The Big Issue Writers.

Shove, E., Watson, M., Hand, M. and J. Ingram (2007) *The Design of Everyday Life.* Oxford and New York: Berg.

Silverman, D. (ed.) (2004) *Qualitative Methods: Theory, Method and Practice.* London: Sage.

Skrbis, Z. (2006) 'Australians in Guantanamo Bay: Gradation of Citizenship and the Politics of Belonging' In N. Yuval-Davis, K. Kannabiran and U. Vieten (eds), *The Situated Politics of Belonging.* London: Sage Publications, pp. 176–190.

SMH. (2006) 'Australia Web Access Running at a Crawl', 9 March, *Sydney Morning Herald,* Accessed 10.04.06, Available at http://www.smh.com.au.

Söderberg, J. (2011) 'Free Space Optics in the Czech Wireless Community: Shedding Some Light on the Role of Normativity for User-Initiated Innovations'. *Science Technology & Human Values* 36:4, 423–450.

Star, S. L. (1999) 'The Ethnography of Infrastructure'. In P. Lyman and N. Wakeford (eds), *Analysing Virtual Societies: New Directions in Methodology, American Behavioural Scientist,* 43:3, 377–391.

Star, S. L. (1995) *Cultures of Computing.* UK: Wiley-Blackwell.

Star, S. L. (1991) 'Power, Technology and the Phenomenology of Conventions: On Being Allergic to Onions'. In J. Law (ed.), *Sociology of Monsters: Essays on Power, Technology and Domination.* London: Routledge.

Street, A. (2011) Artefacts of Not-knowing: The Medial Record, the Diagnosis and the Production of Uncertainty in Papua New Guinean Medicine. *Social Studies of Science,* 41(6): 815–834.

DOI: 10.1057/9781137312532.0015

Thompson, E. (1994) *Fair Enough: Egalitarianism in Australia*. Sydney: University of New South Wales.

Thomson, M. (2007) *Makers, Breakers and Fixers: Inside Australia's Most Resourceful Sheds*. Sydney: Harper Collins Publishers.

Thomson, M. (2006) 'I tinker therefore I am', *Monograph Series 9*, Deer Isle, ME: Haystack Mountain School of Crafts.

Thomson, M. (2002a) *Rare Trades: Making Things by Hand in the Digital Age*. Sydney: Harper Collins Publishers.

Thomson, M. (2002b) *The Complete Blokes & Sheds: Stories from the Shed. Behind the Corrugated-Iron Curtains of Australia's Sheds*. Sydney: Harper Collins Publishers.

Thomson, M. (1999) *Blokes & Barbies: The Legendary Australian Barbeque: Meat, Metal and Fire*. Sydney: Harper Collins Publishers.

Turkle, S. (1995) *Life on the Screen: Identity in the Age of the Internet*. London: Weidenfeld & Nicolson.

Traweek, S. (1988) *Beamtimes and Lifetimes: The World of High Energy Physics*, Cambridge, MA: Harvard University Press.

Turnball, D. (2000) *Masons, Tricksters and Cartographers: Comparative Studies in the Sociology of Scientific and Indigenous Knowledge*. Amsterdam: Hardwood Academic Publishers.

Turnball, S. (2008) 'Mapping the Vast Suburban Tundra; Australian Comedy from Dame Edna to Kath and Kim'. *International Journal of Cultural Studies*, 11:1, 15–32.

van Oost, E., Verhaegh, S. and N. Oudshoorn (2009) 'Innovation Community to Community Innovation User-Initiated Innovation in Wireless Leiden'. *Science Technology & Human Values*, 34:2, 182–205.

Wajcman, J. (2004) *Technofeminism*. Cambridge, MA: Polity Press.

Wajcman, J. (1991) *Feminism Confronts Technology*. Pennsylvania: Pennsylvania State University Press.

Wakeford, N. (2003) 'The Embedding of Local Culture in Global Communication: Independent Internet Cafes in London'. *New Media & Society*, 5:3, 379–399.

Wakeford, N. (1999) 'Gender and the Landscape of Computing in an Internet Café'. In M. Crang and J. May (eds), *Virtual Geographies: Bodies, Space and Relations*. London: Routledge, pp. 178–213.

Warde, A. (2005) 'Consumption and Theories of Practice'. *Journal of Consumer Practice*, 5:2, 131–153.

DOI: 10.1057/9781137312532.0015

Wardell, J. (2013) Australia Election Threatens Shape of $34 Billion Broadband Plan, 15 July, *Reuters*, Accessed 10.10.13, Available at http://www.reuters.com.

Wark, M. (2004) *A Hacker Manifesto*. Cambridge and London: Harvard University Press.

Watson, M. and E. Shove (2008) 'Product, Competence, Project and Practice: DIY and the Dynamics of Craft'. *Journal of Consumer Culture*, 8:1, 69–89.

Weber, K. (1992) 'Imagining Utopia: The Selling of Suburbia'. In *The Lie of the Land*. National Centre for Australian Studies, Melbourne: Monash University.

Whatmore, S. (2003) 'Introduction: Investigating the Field'. In M. Pryke, R. Rose and S. Whatmore (eds), *Using Social Theory: Thinking Through Research*. London: Sage, pp. 67–70.

White, R. (1981) *Inventing Australia: Images and Identity, 1688–1980*. Sydney: Allen & Unwin.

Wyatt, S. (2008) Challenging the Digital Imperative, Inaugural lecture presented upon the acceptance of the Royal Netherlands Academy of Arts and Sciences (KNAW) Extraordinary Chair in Digital Cultures in Development at Maastricht University on 28 March 2008. Accessed at http://www.virtualknowledgestudio.nl

Wyatt, S., Thomas, G. and T. Terranova (2002) 'They Came, They Surfed, They Went Back to the Beach: Conceptualising Use and Non-use of the Internet'. In S. Woolgar (ed.), *Virtual Society? Technology, Cyperbole, Reality*. Oxford: Oxford University Press, pp. 23–40.

DOI: 10.1057/9781137312532.0015

Index

maker culture 4–6, 10–12, 14
making-do 108–109, 112, 123, 130, 134
maps 2, 19, 26, 32, 36, 48, 134
masculinity 32, 52, 108
Men's Shed Movement 51
mess
  digital mess 81–82
  ethnographic mess 34–37, 54
  mess, as in interruption 125, 128
  representational practice 20,
    34–37, 48
methodology, methods. *see also*
  ethnography
  collage photos 32, 36
  installation 32, 48, 96–106
  research website/blog 32
mods, modding, modifications
  culture 127, 130, 134
  material practice 95, 97–99, 102–103,
    108–110
  mundane artefacts, sites 101
  representation 94, 105–107
  sociological practice 37
motley 91
mundane
  cultures and practice 30–31, 79–80,
    122–123
  interruption 59, 64
  sites 21, 33, 82
  technology, infrastructures 14–16,
    130–131

NBN, National Broadband
  Network 6
nodes. *see* antennas

one-size-fits-all 22, 121

participant observation. *see*
  ethnography
portal 54–57

reliability 61, 72
representation
  gender 52, 69
  inscription 20, 46

interruption, instability 59, 67,
  87–88
knowledge 19, 20, 26, 123, 134
mess 36, 48
tinkering, mods 105, 116
WiFi 12, 46–49, 54, 56–57, 69, 92,
  122
raintank 86
rooftops 2–3, 42, 60–61, 63, 71, 83–86

Schwartz-Cowan, Ruth 69
sheds
  DiY 31
  exposed to weather 63
  masculinity 31–32
  part of tech landscape 73, 96, 122
snags. *see* BBQ
social connections 49, 97, 99–100, 108
Star, Susan Leigh 15, 16, 30, 69, 73, 134
sticky tape 94–96, 121–123, 130, 134
STS, Science and Technology
  Studies 4, 15–16, 20, 26, 30, 34, 73
stumbling, site surveying, sweeping the
  sky 41, 77–89, 92
suburbia 27–31, 38, 62, 77, 78, 81–82,
  84, 86

Tech Nest 32
technological imagination 6, 54, 84, 87,
  89–90, 115
The Internet: An Ethnographic Approach
  (Miller and Slater) 17–18, 38,
  109, 129
tinkering
  backyards, local culture 129, 134
  gender 108–109
  mods, modding 95, 98, 102
  never finished 119
  perceived value 113–116
  social connection 107–108
  taking things apart 30, 49
  time 116–117
Tupperware 63, 84
transport
  cycling 33, 97
  car 62–63, 102